Lunar Science - Habitat and Humans

Edited by Yann-Henri Chemin

Published in London, United Kingdom

IntechOpen

Supporting open minds since 2005

Lunar Science - Habitat and Humans
http://dx.doi.org/10.5772/intechopen.95646
Edited by Yann-Henri Chemin

Contributors
Agata Kołodziejczyk, Mateusz Matt Harasymczuk, Fabiana Da Pieve, Natalia E. Koval, Bin Gu, Daniel Muñoz-Santiburcio, Yulia Akisheva, Yves Gourinat, Nicolas Foray, Aidan Cowley, Joseph J. Smulsky, Yann-Henri Chemin

First published in London, United Kingdom, 2022 by IntechOpen
IntechOpen is the global imprint of INTECHOPEN LIMITED, registered in England and Wales, registration number: 11086078, 5 Princes Gate Court, London, SW7 2QJ, United Kingdom
Printed in Croatia

British Library Cataloguing-in-Publication Data
A catalogue record for this book is available from the British Library

Additional hard and PDF copies can be obtained from orders@intechopen.com

Lunar Science - Habitat and Humans
Edited by Yann-Henri Chemin
p. cm.
Print ISBN 978-1-80355-078-7
Online ISBN 978-1-80355-079-4
eBook (PDF) ISBN 978-1-80355-080-0

For EU product safety concerns:
IN TECH d.o.o., Prolaz Marije Krucifikse Kozulić 3, 51000 Rijeka, Croatia, info@intechopen.com or visit our website at intechopen.com.

Meet the editor

Dr. Chemin graduated in Planetary Sciences with Astronomy in 2016, after a full academic path in Earth Observation in Asia. He is now concentrating on Security, Defense and Space Industries (2022), a field filled with both commercial endeavours and highly technical advances. He worked on the hyperspectral analysis of the Apollo 12 landing site and its vicinity on the Moon's surface. He also worked on the thermodynamic energy balance of the surface-atmosphere interactions of the moon Titan and on the mapping of craters on the dwarf planet called Ceres., and keeps a keen eye on challenges to settle on the moon and Mars, especially food creation, settlements construction, and in situ resource utilization. His interests include philosophy, innovation, space science and exploration challenges, spatial sciences, and geopolitical impacts on space.

Contents

Preface

Startups and rocket manufacturers are revolutionizing access to space. For example, Aerojet Rocketdyne, NASA, and SpaceX partially (and increasingly) use 3D-printed elements in rocket building, which lowers costs and allows for faster construction of engines. The number of low Earth orbit (LEO) micro-gravity scientific experiments conducted by the International Space Station is already in the thousands. This year, the first set of circumlunar commercial missions will be launched, opening a vastly more diversified point of view on the possibilities to make the moon our first non-Earth planetary port. Although the multiple challenges of lunar habitats and periodic/long-term crew rotations are still many, science and technology, alongside commercial drives, are changing the convergence rate on a first landing/settlement date.

There are still many open scientific questions about the moon, and there are also now many questions about humans and their potential lives there. This book looks at the history of the moon orbit and the prospects of in situ lunar science, the radiation impact on the lunar surface, the resistance of settlement materials on the moon under the conditions of protecting humans on-site, and the preparation of humans for space missions.

Dr. Yann-Henri Chemin
Joint Research Centre,
Ispra, Italy

Section 1

Introduction

Chapter 1

Access to Space, Access to the Moon – Two Sides of the Same Coin?

Yann-Henri Chemin

Abstract

The dynamics of human expansion towards space are going through Earth external layers, orbital space and the Moon. With its low gravity, slingshot effect relative to Earth, on-site resources and relative proximity to Earth in the solar system, the renewed space race is effectively returning first to the Moon. A psychological bridge to enlarge our civilization with a permanent bridge to our natural satellite. The development of this Earth-Moon system, requires enormous amount of finances, energy, science, technology, but over all, opportunities. This chapter deals with the efforts and the mental changes that may eventually result from all of these changes.

Keywords: space, access to space, access to the Moon, new space race, LEO, Moon, single stage to orbit, near Earth asteroid, astro-mining, ISRU, human health, food security in space, space exploration, Earth-Moon system, human expansion in space

1. Introduction

It is now obvious that the Moon is renewed in its aura of the Earth's first harbor to the solar system and the first non-Earth scientific playground [1], even though it was recognized so more than 60 years ago already [2], alongside the now very reliable Earth orbiting International Space Station (ISS). The Global Exploration Roadmap [3] is a multi-governmental (USA, Canada, Japan, EU, Russia) roadmap to renew space exploration by integrating the Planet Earth in-orbit activities with the Deep Space Gateway in lunar orbit to facilitate sequential and relayed access to both the Moon and Mars, but also potentially the asteroid belt (**Figure 1**).

To foster such a push in space, an accompanying financial support to the economics of space engineering has to be unrolled by the main governments involved in the space race. It turns out that the last few years have seen just that and more. Boosted by the long strides of privately funded commercial ventures to reduce cost to LEO, a new influx of trust sentiment has developed, from both private and public bodies.

This introductory chapter aims at developing the actual state of the '*Access to space*' renewed race. In this highly evolving topic, it will venture into the place of the Moon as *Earth's first spaceport* and as the *solar system exploration gateway*, but also into the *changes of perspective* that need to happen to enable this outward expansion of humanity.

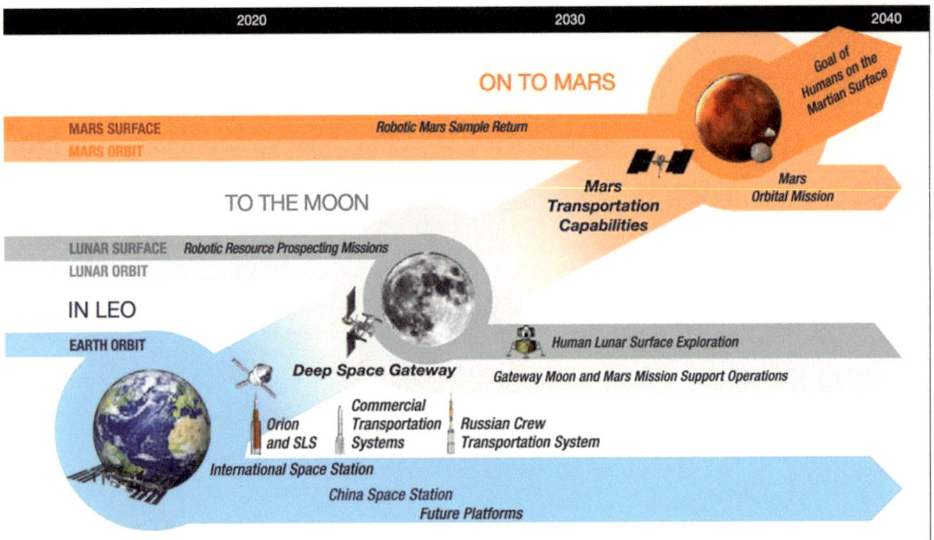

Figure 1.
The global exploration roadmap [3].

2. The increasing level of the investments in space

As a testimony of the recent dynamics of space investments, in 2019 budgets (**Figure 2**), the USA and its now very successful public-private partnerships, are leading the investment into the new space race. For a comparison, in 2019, the French government investment in space was *in par* with Elon Musk private investment.

Fast-foward to 2022, the budgetary update of ESA as reported in the press conference of 13th April 2002, after the 307th ESA HQ meeting, is aiming at negotiating the 2022 Paris agreement with Member States upward of the 14B Euros (15.3B USD), which was the amount of the running Seville Agreement budget. Likewise, the French Government has earmarked 2.6B Euros (2.8B USD; [5]) for 2022 space

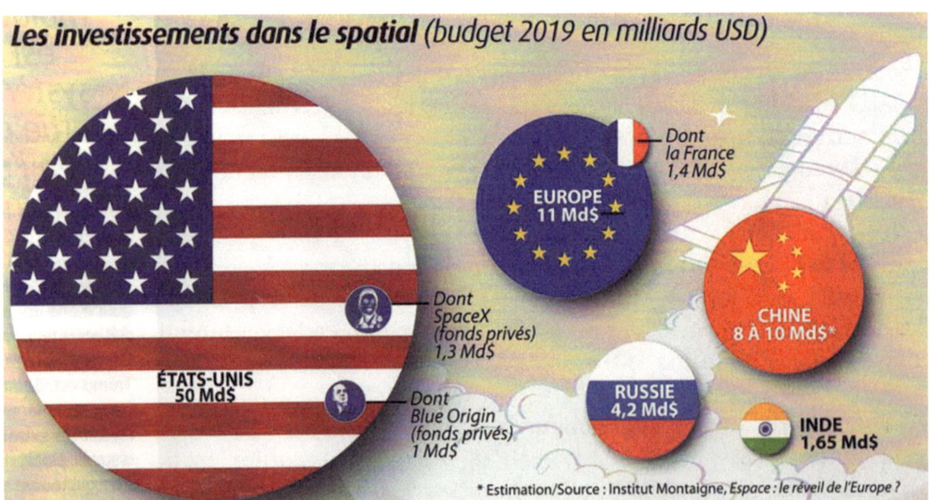

Figure 2.
2019 billion USD investment in the space industry [4].

activities, partially including the very active '*New Space*' sub-sector, which also benefits from additional parallel incubator-type budgets. In the meantime, the NASA FY2022 Request (NASA [6]) is steadily increasing from 22.6B USD in 2020 to 24.8B USD in 2022.

Also worth to mention is that this does not account to the cost of labour and the cost of processed material, in that manner, the PRC's capital/labour efficiency might have some advantage in the medium-term perspectives. Also to observe, the long-lasting experience and multiple collaboration system that Russia and Europe shared in terms of space launching expertise. This was abandoned by the 307th ESA meeting, held on the 13th April 2022, as the direct consequence of the Russian invasion of Ukraine. In the same line and on the same day, the PRC announced that space collaboration with Russia was put on hold (NASA [7]). ESA is now redirecting collaboration efforts primarily with NASA, alongside some other space agencies. In any case, it is not anymore a question of who will get to space; it is more of a question of when and how the upcoming stations and settlements will be kick-started and inhabited.

3. Making a mental shift from using space to being in space

A large part of our effort as a spacefaring society is still within the close vicinity of Earth. As the use of satellites for communications, strategic positioning, global positioning, monitoring of agriculture, climate and environment, among many other applications, are seeing a tremendous break into the daily routines of businesses and policy decision-making, the World looks up to the artificial stars in orbit ever more. This psychological adoption of the orbiting tools, embedded in every vehicle, phone or connected device has made a very long way into making space-borne tools as much ubiquitous as placing a long-distance call or following navigation routes to the regular human. Space is a very real extension of the connected human.

Communication platforms and atomic clocks are in constant connection with human interacting devices. The actual psychological boundary is to make the thought that everyday human physically *can* transfer through that dimension on a more regular basis. To enable such a mental shift to happen practically is a generational effort and a civilizational challenge. It would be a postulate to claim here that we are in the midst of its inception. At the forefront of this is a vision, of many, in their own way, to reach space. There is commercial willingness to follow space as a market, and also a vast amount of engineering exploration, life-long expertise and dedication. The first practical step, the closest, will be orbiting Earth, settling there, making *rendez-vous*, refueling, maintenance, repairs, etc. The second practical step is enabling engineering/commercial solutions to transport *commoners* to/through space according to their needs.

3.1 Enabling engineering of in-orbit transfer, servicing and assembly

In this regard, the Consortium for Execution of Rendezvous and Servicing Operations [8] is a 2016 DARPA-funded industry initiative to create an ecosystem of industries cooperating to generate a long-term viable space industry commercial service environment. The Consortium has eventually morphed in two parts, one focusing on the technical aspect of satellite servicing, initially establishing common standards for safe operations and then, hardware standards for interfaces. On the policy side, preparation is made to inform the regulatory system and to lobby regulators that the Consortium answers properly to their vision.

A satellite servicing platform was launched in 2019 by SpaceLogistics, a subsidiary of Northrop Grumman. The platform is called the Mission Extension Vehicle

(MEV-1). It docked to Intelsat 901 early 2020, a geostationary communications satellite. It took over the attitude and navigation control for the next 5 years of extended operation of the telecommunication satellite. An increment of the concept extends beyond attitude and navigation control. DARPA owned payload (Mission Robotic Vehicle; MRV) will be transported by SpaceLogistics in a future launch (2023) which intends to add on-orbit repair, augmentation, assembly, detailed inspection and relocation of client satellites.

In the same thought process, Orbit Fab, a 2018 start-up, tested in 2019 the on-orbit transfer of water to two pseudo-satellites before transferring the water to the ISS. Called Rapidly Attachable Fluid Transfer Interface (RAFTI), this technology is applicable to fuel, even compressed Xenon [9]. This market segment niche, for the short term, is to support in-orbit relocation services and deployment of pluri-orbit systems. In the longer term, larger opportunities exist with this proven technology to support in-orbit assembly of complex vehicles for long-haul transport in the solar system. The first operational tanker, Tanker-001 Tenzing was launched the 30th of June 2021 by the SpaceX Sherpa-FX orbital transfer vehicle [10].

3.2 Making earth orbit an accessible extension of human life

Like the time revolution that the Concorde provided to cross the Atlantic in the early age of the supersonic civilian flights, a similar revolution could be perceived to go through a certain altitude (stratospheric/sub-orbital, von Karman line, orbital) to fasten travel across opposite sides of Earth. The psychological importance of such a commodity on the human civilization is non-negligeable. Technology mainline acceptance, incremental reliability, robustness, reducing prices, adoption of a way of doing things. In brief, internalization, as a global society, of the extended boundary layer of human daily life. To permit this, the evolving technology should provide extensions of tried and tested means of transport, already widely accepted by all: from an airplane to a spaceplane, there is nearly no psychological barrier when someone climbs in what may look like an airplane, even if it goes much further up, it still may look like what is known to be strongly reliable: an airplane.

Access to LEO by humans on a regular airplane was a long-time revolution to come to fruition. Eventually, evolutions of the 1959 Dyna-Soar [11], the Virgin Galactic Space Ship One [12], the Scaled Composites Model 339 SpaceShipTwo (Scaled Composite, 2010) and the SpaceShipThree released in April 2021 have evolved the spaceplane

Figure 3.
Boeing X20 dyna-soar (1959), virgin galactic StarShipTwo (2010) & three (2021).

DOI: http://dx.doi.org/10.5772/intechopen.105175

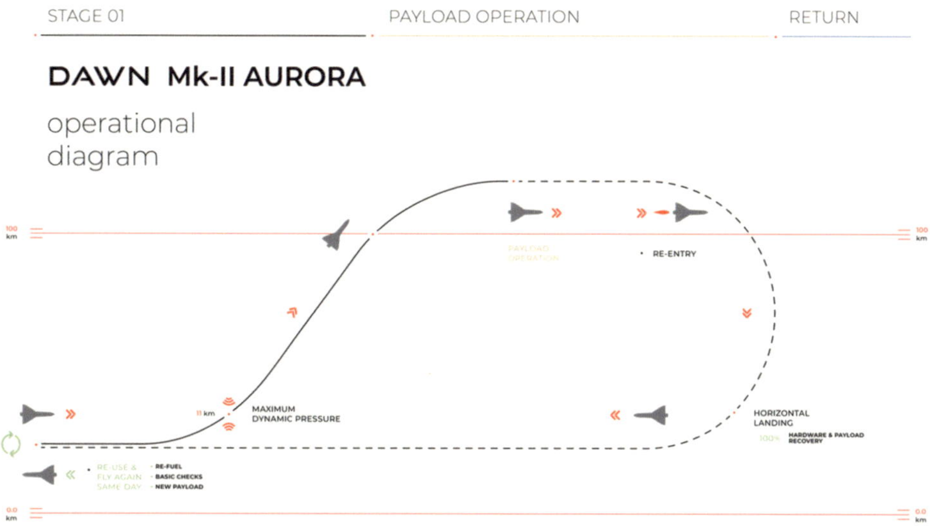

Figure 4.
Operational diagram of Dawn mk-II Aurora single-stage loop to LEO [14].

concept to reduce the cost of oblation protection to much less than the initial design by using the *feather fall* concept (thus the articulated and elongated winglets) that generates a stable high-drag shape for atmospheric re-entry (**Figure 3**). The SpaceShip planes have three flight controls, depending on the in-atmosphere flight being subsonic (manual control) or supersonic (electric) or space (RCS). This space plane launches from larger airplanes (Scaled Composites Model 318 & 348) with a standard runway though longer and larger [13] compared with the Dyna-Soar which was designed smaller and to be launched on top of a rocket.

And yet, closing the single-stage spaceplane surface-LEO-surface loop might just come from New Zealand, where the New Zealand Civil Aviation Authority (CAA) has granted Dawn Aeronautical [14] an Unmanned Aircraft Operator Certificate to fly a suborbital spaceplane from a conventional airport [15]. Dawn aerospace is developing a spaceplane drone (Dawn Mk-II Aurora at this stage of development) in the initial aim at providing a single step to space delivery for smallsat and cubesat-type EO platforms in LEO, with a view to scale the spaceplane design with payload requirements onboard the next iteration, the Mk-III, an 18 metric tons spaceplane. Take-off and landing can be done (**Figure 4**) from any standard runway in the world, permitting government and private companies to have at hand the launching platform and insert the payload directly in their premises. In all, spaceplanes, whether airborne launched/runway landing (Virgin Galactic) or runway take-off/landing (Dawn Aerospace) is decreasing vastly costs to LEO for both passengers and freight. But the most important part is the *ubiquity of launching geolocations* that they address, vastly reducing some of the most common risks and costs to space payloads. Also, bringing spaceplanes one step closer to be visible to all in daily life, a critical part of society adoption of the extended space dimension these carry within.

At this point, we, as a people, reach the very close to Earth external boundary and orbital space. A vastly non-negligeable step in itself to appreciate the civilizational expansion as a daily happening. As with all progresses and development, many steps are intertwined, as it is in these years, both sub-orbital, orbital and the next part to be focussed on. That part is one step further, more difficult, more careful, in terms of engineering, more complex to survive as a human without increasing amount of assistance. Further away, yet also orbitally intertwined with us.

4. Going further, expanding earth to an earth-moon system

4.1 Moon segment: in-orbit station and missions, surface research station

The SpaceX Crew Dragon mission Inspiration4 [16], launched on 16th September 2021 and was sponsored by Entrepreneur Jared Isaacman, acting as mission commander with a crew of three astronauts (Prof./Pilot Sian Proctor, Medical Officer Halley Arceneaux, Mission Specialist Christopher Sembroski). Followed on the 30th of March 2022 by the Axiom sponsored fully civilian crew visiting the ISS for few days onboard Crew Dragon [17]. The SpaceX Starship is a multi-use Moon/Mars transport/settlement vehicle, with return capability. It is now in very fast pace development with early tests proving automatic attitude controls for lifting, landing and free-falling. On the 15th of May 2021, Starship Number 15 (SN15) proved the autonomous Vertical Take Off and Landing (VTOL) capability of the spacecraft. The actual development is the stacking of SN20 above the Super Heavy Booster (**Figure 5**) to permit Earth escape velocity. One of the first use of the starship will be the space tourism *#dearMoon* project (dear [19]), sponsored by the Japanese Entrepreneur Yusaku Maezawa, with the aim at inspiring a broad range of artists by having them do a fly-by of the Moon by 2023. This will also bring SpaceX in the forefront of the Moon transport industry. In parallel, plans were for a first Martian crew by 2024 [18], but eventually are going to be delayed a few years until circumlunar flights of starship become a viable resource to SpaceX.

The Artemis program [20] is a Moon landing and settlement mission from the USA. It is aimed at using public-private partnership to access the Moon in the years ahead and eventually create a Moon's south pole settlement. In preparation, Artemis 2 is planned to be the first crewed mission to perform a flyby of the Moon. To reach to the Moon surface, but also as a relay to Mars, the Deep Space Gateway (**Figure 6**: [3]) also called the Lunar Gateway [21] is created as a relay station, and lunar in-orbit space port. Canada, EU and Japan space agencies have formalized partnerships for the Lunar in-orbit station [21]. The *lunar Gateway* is also clearly defined as a relay station en-route to Mars. On 16th April 2021, NASA granted 2.9B USD to SpaceX to create the ARTEMIS lunar lander, from its Starship series of spacecraft.

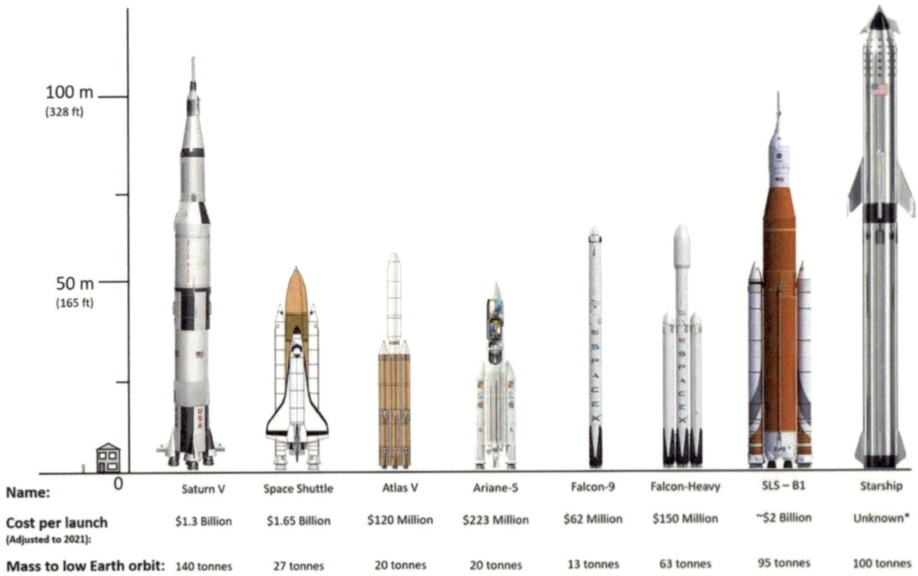

Figure 5.
NASA's SLS and SpaceX's Starship [18].

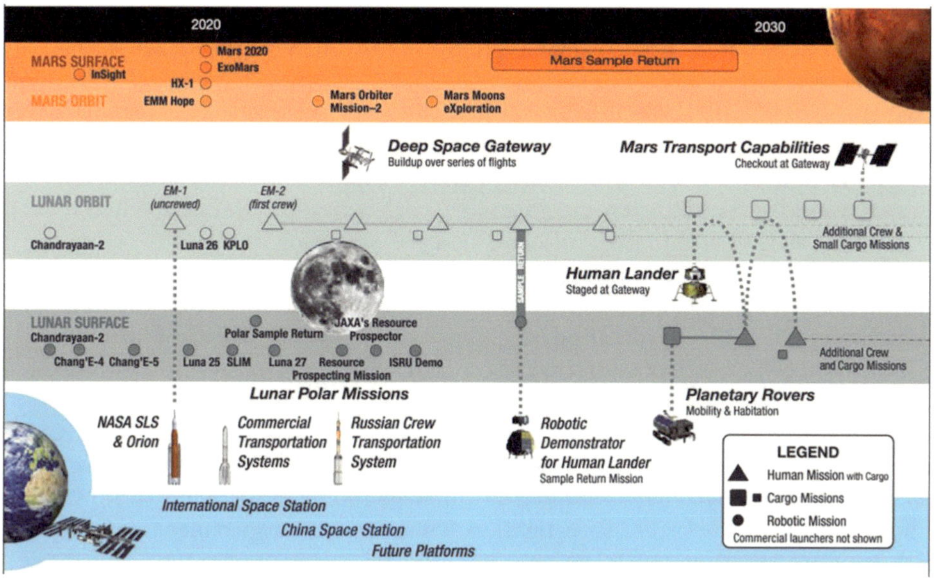

Figure 6.
Government's missions timeline [3].

With the successful lunar orbit injection of Chandrayaan-2 in 2019, the Indian government approved a follow-on mission, Chandrayaan-3, which is launching in 2023 [22], carrying both a lander and rover, with the same set of payloads as Chandrayaan-2, but without an orbiter (C-2 has already successfully inserted an orbiter). ISRO is also working with JAXA to conduct a feasibility study for a joint lunar exploration mission.

The PRC has now fully and officially, engaged into the race to create a Moon base. Weiren [23] announced that the PRC's '*next steps in [...] lunar exploration endeavour will be challenging and demanding as we aim to set up a scientific outpost on the moon's south pole. In the near future, we will also send our astronauts to land on the moon*'. The station will aim at an international cooperation, named the International Lunar Research Station (ILRS). On 21st February 2021, Roscosmos confirmed that the administrative steps to join the ILRS are ready to be signed with the PRC [24]. Across the board, the targets are 2030 for robotic and short-term human presence. By 2036, ILRS settlement and long-term human presence are expected [25]. Prior to the Ukraine invasion, and the subsequent EU sanctions, both EU and Russia had explored the collaboration possibilities, by May 2022, the time of reviewing this text, there is a certain level of uncertainty about the EU positioning on that subject.[1]

4.2 Human health, its requirements and limitations in space and on the Moon

Surface/elevation mapping, exosphere, radiation and volatile composition are the main subjects of interest for governments involved in lunar science, besides the engineering proof of work for orbital control, landing automation and sample return. Its lack of atmosphere also brings advantages to astronomical observation, more so with a large distributed array [26]. The Moon surface is in itself a human health issue, from the micro particulate size of the Moon regolith and its toxicity

[1] At the time of the editing of this manuscript (May 2022), the state of the Russian-lead presence in space is undefined and will certainly be delayed on many planned ventures. The 307th ESA HQ meeting press conference is found here (https://www.youtube.com/watch?v=P1Yox7Jmzlk) with elements provided on stopping all collaborations with Russia.

[27]. Indeed, the lunar environment has detrimental impacts on the human organism [28], but as found in the Global Exploration Roadmap (**Figure** 7; [29]), the Moon environment and astronauts in transits have much less to fear on the Moon base than in deep space or on a 3-year return mission to Mars. Here, we note in **Figure** 7 about fractional Earth gravity that the impact of even a fraction of Earth gravity is highly beneficial on the human health. Indeed, only 1/10–1/2 of Earth gravity would suffice to bring considerable health benefits to humans, whether on the Moon, Mars or within the artificial gravity of a rotating spacecraft.

Most of the orbiting space stations study human health and adaptation to micro-gravity. PRC's TianHe core module launched end of April 2021, the station's environment is studying mutation breeding, medicines and new materials. After 6 months in orbit, a taikonaut returned to Earth 18th May 2022, enabling study of the effect of micro-gravity on the longest period in space for a woman.

Similarly, the ISS is running a large number of scientific experiments on a rotating basis, recently, the Dragon Crew-1 had a Multi-purpose Variable-G Platform (MVP) Cell-06 experiment dedicated to Cartilage–Bone–Synovium (CBS) Micro-Physiological System (MPS) Investigation. This particular experiment is vital to low-gravity adaptation of CBS system. It is especially dealing with the problematic recovery from trauma that happened in space or on the Moon due to its micro-gravity compared to our Earth-adapted bodies (**Figure 8**).

4.3 Human protection and primary survival needs in space and on the moon

The main human protection requirement in space and on the surface of the Moon (which has technically no atmosphere and no magnetic field) is radiation

Main Human Health and Performance Risks for Exploration	Mission: ISS (6 mo)	Lunar (6 mo)	Deep Space (1 yr)	Mars (3 yr)
Legend: Not mission limiting – GO (green); Not mission limiting, but increased risk – GO (yellow); Mission limiting – NO GO (red)				
Musculoskeletal: Long-term health risk of early onset osteoporosis Mission risk of reduced muscle strength and aerobic capacity	Green	Green	Yellow	Red
Sensorimotor: Mission risk of sensory changes/dysfunctions	Green	Yellow	Green	Yellow
Ocular Syndrome: Mission and long-term health risk of microgravity-induced visual impairment and/or elevated intracranial pressure	Yellow	Yellow	Red	Red
Nutrition: Mission risk of behavioral and nutritional health due to inability to provide appropriate quantity, quality and variety of food	Green	Green	Yellow	Red
Autonomous Medical Care: Mission and long-term health risk due to inability to provide adequate medical care throughout the mission (Includes onboard training, diagnosis, treatment, and presence/absence of onboard physician)	Green	Yellow	Red	Red
Behavioral Health and Performance: Mission and long-term behavioral health risk	Green	Yellow	Red	Red
Radiation: Long-term risk of carcinogenesis and degenerative tissue disease due to radiation exposure–Largely addressed with ground-based research	Green	Yellow	Red	Red
Toxicity: Mission risk of exposure to a toxic environment without adequate monitoring, warning systems or understanding of potential toxicity (dust, chemicals, infectious agents)	Yellow	Yellow	Yellow	Red
Autonomous Emergency Response: Medical risks due to life support system failure and other emergencies (fire, depressurization, toxic atmosphere, etc.), crew rescue scenarios	Yellow	Yellow	Red	Red
Hypogravity: Long-term risk associated with adaptation during intravehicular activity and extravehicular activity on the Moon, asteroids, Mars (vestibular and performance dysfunctions) and postflight rehabilitation	Green	Green	Yellow	Red

Figure 7.
Health concerns in space (from the [29]) Health concerns Legend, Red (Unacceptable): A risk with one or more of its attributes (i.e. consequence, likelihood, uncertainty) currently exceeding established human health and performance standards for that mission scenario. Yellow (Acceptable): A risk with all of its attributes (i.e. consequence, likelihood, uncertainty) well understood and characterized, such that they meet existing standards but are not fully controlled, resulting in 'acceptance' of a higher risk posture. Lowering the risk posture is important, but the risk is not expected to preclude a mission. Green (Controlled): A risk with all of its attributes (i.e. consequence, likelihood, uncertainty) well understood and characterized, with an accepted mitigation strategy in place to control the risk. It is still helpful to pursue optimized mitigation opportunities such as compact and reliable exercise devices.

Figure 8.
Glover installing MVP Cell-06 in the ISS (courtesy: NASA spaceflight Center).

protection. In this matter, another set of experiments on Cygnus transport aims directly at the Moon. The Artemis HERA on Space Station (A-HoSS) modifies the Hybrid Electronic Radiation Assessor (HERA), built to operate as the primary radiation detection system for Orion and certified for flight on Artemis 2, to operate on the space station. The investigation provides an opportunity to evaluate this hardware in the space radiation environment prior to the Artemis 2 flight [30].

Radiation is both a human and a technological constraint. Indeed, shielding for radiation takes weight and redundancy for technology. Humans on the other hand, are subject to certain radiation related diseases, and shielding them, as well as making them more resistant to regular radiation doses, is still the centre of many designs, experiments and research works. On the Moon, a most ubiquitous radiation protection is the use of regolith above settlements' habitats. This would serve a combined purpose of mechincal kinetic absorption for micro-meteorids and radiation shielding.

The main human survival need *is* oxygen (counted in 1–3 minutes roughly), then it *is* water (counted in 2–5 days roughly). Recycling water is altogether a strategic endeavor in many ways, as water redundancy is very costly in weight to space, and recycling it increases safety and resilience of space explorers and dwellers. As shown by the Martian rover Perseverance soon after its arrival, ISRU, in this case the CO_2 transformed in O_2, is the most economic and reliable way to have access to non-limited resources once on a given planetoid surface. While O_2 can be easily compressed for transport, and uncompressed into a globally controlled environment system in some inhabitations on another planetoid, it is not the same for H_2O, as it is incompressible, and vastly more dense than O_2. Finally, H_2O can be split in H_2 and O_2, providing respectively energy and life support if needed.

4.4 Why a lunar station at the south pole: Aitken basin?

On the Moon surface, a practical ISRU is water extraction, which can only be found in the vicinity of the poles, where some craters bottom are always under shadow, keeping frozen ice from sublimating [31]. This is where (the South Pole – Aitken basin) the joint Sino-Russian International Lunar Research Station was to be built in the coming years (to be redefined the after May 2022 hold of PRC from Russian space ventures; NASA [7]), for the obvious reason of water ISRU (**Figure 9**). Nevertheless, water recycling technology will always be a must on the surface of the Moon, first of all, for

Figure 9.
Chandrayaan-1 M3 lunar surface water. South & North poles [31].

emergency reasons, but also for the settlement missions when ISRU water mining/extraction tools will be on inititialization/maintenance phases and return only partial yield.

4.5 Food security, ISRU and fluxes circular sustainability

Already 60+ years ago, Trudeau [2] mentioned that settlements in man-made tubes underground the Moon regolith would provide mechanical protection from meteoroids gardening and cosmic radiation. He also added that '*Early attention will be given to hydroponic culture of salads and the development of closed cycle food production systems*'. In 2021, NASA and CSA launched the Deep Space Food Challenge [32], to increase innovation in the field of nutrition and make the food tasty to '*encourage people to continue eating during long space voyages*', recognizing the psychological impact of appearance, something understood by earth livestock supplement producers early on. Additionally, the challenge aims at finding new food production technologies (or systems), which would have little waste produced or resources required.

'*ExoAgriculture*', a term coming from the exogenically designed and applied agriculture, as opposed to the Earth-based endogenic agriculture, is a new field of science in itself (also referred to as *ExoAg*), and is being developed now constantly in the micro-gravity of the ISS, alongside being experimented on various Martian soil analogs in various research laboratories on Earth [33]. As for the Moon, in the absence of atmosphere to deflect/absorb both cosmic radiation and some of the meteoroids, the actual constant and thorough gardening of the regolith has led to the reduction of its differentiation and down-homogeneization of the mineralogical properties. The result being powder/dust with less to no macro-structure or molecular properties, radiation further sterilized its macro-properties by breaking molecular loose crystalline structures. Even if some Moon rocks maybe mechanically crushed into structurally interesting macro structured soil interesting to plants root development, it will most probably still require additional minerals, Earth microbes and fertilizers to grow on it, returning to an initial perspective from 1959 mentioned above which might be more practical.

Besides the most prominent human survivability performance improvements by the benefits of adding vitamins, minerals to the diet, as well as recycling CO_2 into O_2 as a redundancy of mechanical/chemical scrubbers, the actual proximity to and care of plants may also affect positively the long-term conditions of life of humans in adverse environments, though a reviews calls for more experiments on this [34]. Yet, the probability of adding a plant or a frame with a natural environment in a windowless office has been found five times more attractive than not adding any in a more recent study from Bringslimark et al. [35].

5. Moon commercial activities and advanced ISRU

5.1 Space mining laws on earth

'*Space mining*' as a concept was born out of the 1990s. It practically kick-started when President Obama signed what is now coined as the '*Space Law*' on November 25th, 2015. It was later ratified by the US Congress, giving it an independent life. The fast pacing development of private-public partnerships to reach space is reaching an ever reducing cost and repetition of launch per week by the same launchers (SpaceX), and even now the development of suborbital launchers from standard airplanes (Virgin Orbit), lastly one stage airplane to suborbital apogee (Dawn). Companies with clear interest in asteroid mining are among many others: Deep Space Industries, Orbital Sciences Corporation, Mars One, Bigelow Aerospace, etc.

In 2016, the Grand Duchy of Luxembourg announced the creation of Space Resources, an initiative to generate a long term investment in space resources extraction and use. Initially received with mixed belief and with a concentrating focus on astromining, it nonetheless attracted 50+ companies with a mandate to do astromining. One of the most impressive gesture it had, was to sign the 1967 Outer Space Treaty, later (October 2020) reinforced by the Artemis Treaty. Luxembourg, in 2017, modified its own constitutional laws to allow local companies to exploit space resources. At the time of the writing of this chapter [36], the Grand Duchy is joining NASA in the hope to develop businesses that can provide Artemis Base Camp with extraction and processing of vital in situ resources like oxygen and fuel (H^3, Thorium, etc.).

5.2 Near moon commercial activities example: near-earth-asteroid mining

The Near-Earth-Asteroids (NEAs) approaching Earth are monitored by NASA [37], their weekly entry into the 0.05 AU Earth boundary are recorded and accounted for. *Several NEOs per week are found, with diameters reaching 100+ meters for some!* The common content of such asteroid in *high-quality* Fe, Ni, Pt, Co, etc. makes them attractive astromining targets in term of energy spent to reach a resource. Considering that 0.05 AU is about 7.5 Millions of km, and the Moon is at 0.4 Millions of km from Earth, with its low gravity and slingshot effect relative to Earth. There is energy-based reason to consider the Moon as a NEA mining base. Blair [38] already identified the impact of NEO asteroid mining on the Platinum market and Earth supply streams, LL chondrites already found in NEOs are indeed rich in Pt, at a rate of 32 + k USD per kg, the economics of reaching to space for astromining are getting vastly efficient as time passes and Earth mining pressure increases.

As an example, the NEO asteroid 2011 UW 158 (**Figure 10**) made the news in July 2015 when it passed near Earth at 2.4 millions of km. It is 452 m x 1011 m and

Figure 10.
NEA 2011UW 158, carrying 90 million tons of platinum [39].

was estimated to carry 90 Millions of tons of Platinum (2022 value: 2.9 Trillions of USD). In recent news, JPL jointly with NASA announced the completion of the payload assembly in April 2021 for a mission to Psyche, a peculiar asteroid since it exhibits all the characteristics of the remnants of a nickel-iron rocky planet core ([40]; NASA [41]). Here again besides the obvious scientific treasure trove about inner rocky planet origins, formation, geomagnetism and actual composition, there is the question of the purity of the ore found there, and the potential for in situ extraction or redirection to another location for exploitation.

As an example of advanced ISRU, metal 3D-printing technology matures in micro-gravity (i.e. [42]), with its Lithography-based Metal Manufacturing (LMM), it is rather straightforward to envisage the use of high-quality metal ore into such machine, in space or on a planetoid, for infrastructure creation, whether spacefaring designs or settlement on-the-spot adapted constructions. Obvious questions of heating power source will come to mind, but there are already such solutions rolling on six wheels on the surface of Mars, it would require some scaling, but it is in the realm of feasible and transportable.

5.3 Optimizing economic returns and ISRU of NEAs

Literally, any rock flying around is an addition to Earth resources if safely added to the Earth surface. Likewise, it can become an addition to any rocky moon or planet in its economical action range. In this regard, Vergaaij et al. [43] have economically optimized orbits and trajectories to maximize both ISRUs and economic returns in the Earth-Moon-Mars settlement ecosystem (**Figure 11**).

In the context of this contribution, it becomes logical that the Earth-Moon system, and the Moon as a springboard to low-energy access to local space permits vast economies of transport from and to NEAs. Whether pre-processing can be done in space, on the Moon surface, are still to be evaluated. It is in the author's belief that 'technically' some level of metal 3D printing should be

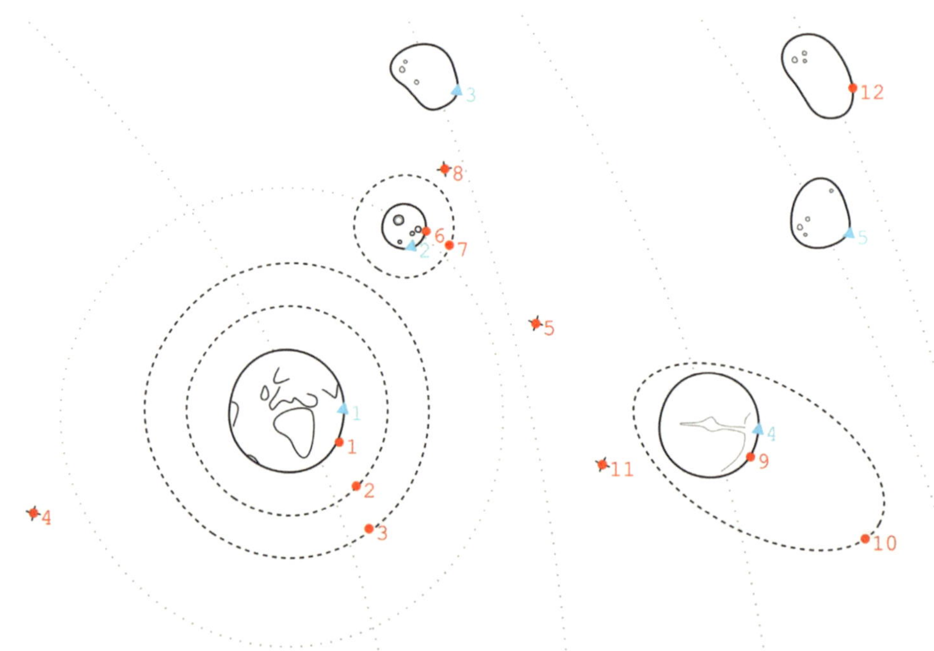

Figure 11.
Economic ecosystem of space resources in earth vicinity [43].

feasible on the Moon surface very soon, as for a timeline for economic viability, it is still an unknown.

Notes: Blue triangles: resource locations (1) Earth surface, (2) Lunar surface, (3) NEA, (4) Martian surface, and (5) MBA. Red circles: customer locations (1) Earth surface, (2) LEO, (3) GEO, (4) SE-L1, (5) SE-L2, (6) Lunar surface, (7) LLO, (8) Lunar Gateway, (9) Martian surface, (10) Mars Base Camp, (11) SM-L1 and (12) Psyche.

6. Conclusions

In a few years, space resources, space economics and the space race in general have had a major boost from private-public partnerships, as well as from the space narratives of (now many) Earth governmental entities. The now nearly weekly launch or test of SpaceX rockets, the ever-growing number of space related start-ups, the various economic pressures they generate on well-established slow R&D turn over space mammoths is as much a breath of fresh air that it is a renewed psychological return to space for humanity.

While research on human in space and microgravity is on a full swing in ISS, humans' adaptation to planetoids/moons of lower gravity than Earth can be fast-forwarded on the Moon alongside many more practical research and logistical developments about actually living on a planetoid/moon.

The space ecosystem for humans' explorers, resources acquisition and settlements requirements are all identified. The first part of the Chinese Earth orbital space station is in LEO for several months already. The ISS has the largest crew on-board for a long time, increasing the space experience of more people at the same time. The first fully commercial crew flying and staying onboard ISS is the Axiom Ax-1 crew (shuttled by a SpaceX Dragon-Crew capsule) in April 2022. The commercial space station *Orbital Reef* is now is early phases (Orbital [44]), the *Axiom* space station too [45].

The NASA Moon orbital Gateway to Mars, the PRC-lead International Lunar Research Station, are all effectively getting built. SpaceX is sending tourists around the Moon (Yusaku Maezawa and 10–12 artists) by 2023 on a 6 days trip to lunar circumvolution. NASA selected SpaceX to land people on the Moon initially by 2023, then by 2025, thought it might still take more time (NASA [46]), NASA issued a second round of call in March 2022 for other commercial Moon landing options, reaching the Moon surface by an initially stated target of 2027 [47].

The Moon is *de facto* a gravity well for our expanding civilization and its willingness to enlarge its dimension of economics of resources, as a proxy of developmental support. As importantly, it enravels a primordial corollary, making us, humans, an Earth-Moon species. Shall we?

Acknowledgements

The author would like to thank the anonymous reviewer for the insightful comments permitting to reinforce the central claims and improve the quality of the manuscript.

Abbreviations

A-HoSS	Artemis HERA on Space Station
ARTEMIS	Moon landing and settlement program of NASA

CONFERS	Consortium for Execution of Rendezvous and Servicing Operations
CAA	Civil Aviation Authority
DARPA	Defense Advanced Research Projects Agency
EO	Earth Observation
GATEWAY	Lunar Orbit transfer station of NASA
GEO	Geostationary Earth Orbit
GER	Global Exploration Roadmap
HERA	Hybrid Electronic Radiation Assessor
HORIZON	Moon Base project of the US Army in 1959
ILRS	International Lunar Research Station
ISS	International Space Station
ISRO	Indian Space Research Organization
ISRU	In-Situ Resources Utilization
JAXA	Japan Aerospace Exploration Agency
LEO	Low Earth Orbit
LLO	Low Lunar Orbit
LMM	Lithography-based Metal Manufacturing
MBA	Main Belt Asteroid
MEV	Mission Extension Vehicle
MRV	Mission Robotic Vehicle
MVP	Multi-purpose Variable-G Platform
NASA	National Aeronautical and Space Agency
NEA	Near-Earth Asteroid
NEO	Near-Earth Objects
PRC	People's Republic of China
PRAC	Porto Rico Arecibo Observatory
RAFTI	Rapidly Attachable Fluid Transfer Interface
RCS	Reaction Control System
SM-L1	Sun-Mars Lagrangian 1
USA	United States of America

Author details

Yann-Henri Chemin
Joint Research Centre, Ispra, Italy

*Address all correspondence to: dr.yann.chemin@gmail.com

References

[1] Chemin YH. Introductory Chapter: A Tipping Point for a Return to the Moon. In: Lunar Science. Croatia: IntechOpen; 2019. p. 1

[2] Trudeau AG. Project Horizon, a Moon Base. Declassified US Army material by authority of form 1575 dtd 21 Sep 1961 by Lt. Col. Donald E. Simon. CS; 1959

[3] GER. The Global Exploration Roadmap, Version 2018 [Internet]. 2018. Available from: https://www.nasa.gov/sites/default/files/atoms/files/ger_2018_small_mobile.pdf [Accessed: February 1, 2021]

[4] Diplomatie. Vers une nouvelle course a l'espace. Les Grands Dossiers No 58. Diplomatie GD, Affaires Strategiques et Relations Internationales. USA, Chine, Russie, Europe: Areion. news, October-Novembre 2020; 2020 Available from: https://www.areion24.news/produit/les-grands-dossiers-de-diplomatie-n-58/

[5] CNES. CNES in 2022, 60 Years Serving French and European Space Sights Firmly Set on New Spaces [Internet]. 2022. Available from: https://presse.cnes.fr/en/cnes-2022-60-years-serving-french-and-european-space-sights-firmly-set-new-spaces [Accessed: April 14, 2022]

[6] NASA Budget. Financial Year 2022 Budget Request [Internet]. 2022. Available from: https://www.nasa.gov/sites/default/files/atoms/files/fy2022_budget_summary.pdf [Accessed: April 14, 2022]

[7] NASA Watch. China Is Halting Science Cooperation with Russia [Internet]. 2022. Available from: http://nasawatch.com/archives/2022/04/china-is-haltin.html [Accessed: April 15, 2022]

[8] CONFERS. Consortium for Execution of Rendezvous and Servicing Operations (CONFERS) [Internet]. 2016. Available from: https://www.ati.org/collaborations/confers [Accessed: February 1, 2021]

[9] Mayfield M. Industry offering on-orbit satellite servicing. National Defense Magazine [Internet]. 2021. Available from: https://www.nationaldefensemagazine.org/articles/2021/1/29/industry-offering-on-orbit-satellite-servicing [Accessed: February 1, 2021]

[10] Werner D. Orbit Fab to Launch First Fuel Tanker in 2021 with Spaceflight [Internet]. 2020. Available from: https://spacenews.com/orbit-fab-to-launch-with-spaceflight/ [Accessed: February 1, 2021]

[11] Boeing. 2021. Available from: https://www.boeing.com/history/products/x-20-dyna-soar.page [Accessed: February 17, 2021]

[12] Scaled Composites. Model 316: Space Ship One [Internet]. 2003. Available from: https://www.scaled.com/portfolio/spaceshipone/ [Accessed: February 17, 2021]

[13] Space Port America. Press kit [Internet]. 2021. Available from: https://www.spaceportamerica.com/wp-content/uploads/2020/11/Press-Kit-Nov-1.pdf [Accessed: February 2, 2021]

[14] DAWN. DAWN Aerospace Website [Internet]. 2021. Available from: https://www.dawnaerospace.com/ [Accessed: February 17, 2021]

[15] Messier. Dawn Aerospace Licensed to Fly New Zealand's First Spaceplane [Internet]. 2020. Available from: http://www.parabolicarc.com/2020/12/09/dawn-aerospace-licensed-to-fly-new-zealands-first-spaceplane/ [Accessed: February 17, 2021]

[16] Stimac V. Space X Announces First all-Civilian Mission to Space, Inspiration [Internet]. 2021. Available from: https://www.forbes.com/sites/valeriestimac/2021/02/01/spacex-announces-first-all-civilian-mission-to-space-inspiration4/ [Accessed: February 3, 2021]

[17] Stimac V. Axiom Space Announces Ax1 – First all-Private Crew to Visit ISS [Internet]. 2021. Available from: https://www.forbes.com/sites/valeriestimac/2021/01/26/axiom-space-announces-ax1--first-all-private-crew-to-visit-iss/ [Accessed: February 3, 2021]

[18] Dorian, G., Whittaker, I., 2021. Space X vs Nasa: Who will get us to the Moon first? Here's how their latest rockets compare. Available from: https://theconversation.com/spacex-vs-nasa-who-will-get-us-to-the-moon-first-heres-how-their-latest-rockets-compare-154199

[19] Dear Moon. Main webpage [Internet]. 2021. Available from: https://dearmoon.earth/ [Accessed: February 1, 2021]

[20] ARTEMIS. The Artemis Program Web Page at NASA [Internet]. 2020. Available from: https://www.nasa.gov/artemisprogram [Accessed: February 1, 2021]

[21] GATEWAY. In Lunar Orbit, Gateway System and Partnerships [Internet]. 2020. Available from: https://www.nasa.gov/gateway [Accessed: February 1, 2021]

[22] TOI. Two Gaganyaan Abort Tests in August, December; Relay Satellites next Year [Internet]. 2022. Available from: https://timesofindia.indiatimes.com/india/2-gaganyaan-abort-tests-in-august-december-relay-satellites-next-year/articleshow/90688885.cms [Accessed: April 20, 2022]

[23] Lei Z. Nation Planning Scientific Station on the Moon [Internet]. 2020. Available from: http://www.chinadaily.com.cn/a/202009/09/WS5f581650a310675eafc585c4.html [Accessed: February 1, 2021]

[24] Jones A. Russia, China to Sign Agreement on International Lunar Research Station. Space News Website [Internet]. 2021. Available from: https://spacenews.com/russia-china-to-sign-agreement-on-international-lunar-research-station/ [Accessed: February 18, 2021]

[25] Zakarin J. China Makes Secret Spacecraft and Moon Landing Moves to Heat up Space Race [Internet]. 2020. Available from: https://observer.com/2020/09/chinas-moon-base-advances-and-secret-spacecraft-heat-up-the-new-space-race/ [Accessed: February 1, 2021]

[26] Crawford IA, Anand M, Cockell CS, Falcke H, Green DA, Jaumann R, et al. Back to the Moon: The scientific rationale for resuming lunar surface exploration. Planetary and Space Science. 2012;**74**(1):3-14

[27] Linnarsson D, Carpenter J, Fubini B, Gerde P, Karlsson LL, Loftus DJ, et al. Toxicity of lunar dust. Planetary and Space Science. 2012;**74**(1):57-71

[28] Khan-Mayberry N. The lunar environment: Determining the health effects of exposure to moon dusts. Acta Astronautica. 2008;**63**(7-10):1006-1014

[29] GER. The Global Exploration Roadmap, Version 2013 [Internet]. 2013. Available from: https://www.nasa.gov/sites/default/files/files/GER-2013_Small.pdf [Accessed: February 1, 2021]

[30] NASA Science Highlights. A-HoSS: Artemis HERA on the Space Station [Internet]. 2021. Available from: https://www.nasa.gov/mission_pages/station/research/news/ng-15-science-highlights [Accessed: April 27, 2021]

[31] Pieters CM, Goswami JN, Clark RN, Annadurai M, Boardman J, Buratti B,

et al. Character and spatial distribution of OH/H_2O on the surface of the Moon seen by M3 on Chandrayaan-1. Science. 2009;**326**(5952):568-572

[32] DSFC. The Deep Space Food Challenge [Internet]. 2021. Available from: https://www.deepspacefood challenge.org/ [Accessed: April 3, 2021]

[33] Fackrell LE, Schroeder PA, Thompson A, Stockstill-Cahill K, Hibbitts CA. Development of martian regolith and bedrock simulants: Potential and limitations of martian regolith as an in-situ resource. Icarus. 2021;**354**:114055

[34] Bringslimark T, Hartig T, Patil GG. The psychological benefits of indoor plants: A critical review of the experimental literature. Journal of Environmental Psychology. 2009;**29**(4):422-433

[35] Bringslimark T, Hartig T, Grindal Patil G. Adaptation to Windowlessness: Do Office workers compensate for a lack of visual access to the outdoors? Environment and Behavior. 2011;**43**(4):469-487

[36] FT. Luxewmbourg Space Programme to Work with Nasa on Moon Mining. The Financial Times [Internet]. 2021. Available from: https://www.ft.com/content/3ced3460-abf2-4048-bce4-66f01e16ade4 [Accessed: February 17, 2021]

[37] NASA. Center for near Earth Object Studies, near Earth Close Approaches [Internet]. 2021. Available from: https://cneos.jpl.nasa.gov/ca/ [Accessed: February 4, 2021]

[38] Blair BR. The role of near-earth asteroids in long-term platinum supply. Space Resources Roundtable. 2000;**II**:1070

[39] PRAO. Puerto Rico's Arecibo Observatory Provides the First Detailed Images of Asteroid 2011 UW158 [Internet]. 2015. Available from: https://twitter.com/PuertoRicoPUR/status/622219032263921667 [Accessed: February 4, 2021]

[40] JPL. The Psyche Mission [Internet]. 2021. Available from: https://www.jpl.nasa.gov/missions/psyche [Accessed: April 3, 2021]

[41] NASA Psyche. NASA Mission to Psyche [Internet]. 2021. Available from: https://www.nasa.gov/psyche [Accessed: April 3, 2021]

[42] Incus. New Project to Test Incus Technology for 3D-Printing Spare Parts in Space [Internet]. 2022. Available from: https://www.incus3d.com/news-entry/view_express_entity/342 [Accessed: April 15, 2022]

[43] Vergaaij M, McInnes CR, Ceriotti M. Comparison of material sources and customer locations for commercial space resource utilization. Acta Astronautica. 2021;**184**:23-34

[44] Orbital Reef. The Orbital Reef Space Station [Internet]. 2022. Available from: https://www.orbitalreef.com/ [Accessed: April 20, 2022]

[45] Axiom. The Axiom Space Station [Internet]. 2022. Available from: https://www.axiomspace.com/axiom-station [Accessed: April 20, 2022]

[46] NASA Artemis. NASA'S Management of the Artemis Missions. NASA Office of Audits [Internet]. 2021. Available from: https://oig.nasa.gov/docs/IG-22-003.pdf [Accessed: April 20, 2022]

[47] NASA Second Call. NASA Provides Update to Astronaut Moon Lander Plans under Artemis [Internet]. 2022. Available from: https://www.nasa.gov/press-release/nasa-provides-update-to-astronaut-moon-lander-plans-under-artemis [Accessed: April 20, 2022]

Section 2

Lunar Science

Chapter 2

The Evolution of the Moon's Orbit Over 100 Million Years and Prospects for the Research in the Moon

Joseph J. Smulsky

Abstract

As a result of solving the problem of interaction of Solar-system bodies, data on the evolution of the Moon's orbit were obtained. These data were used as the basis for the development of a mathematical model for the Moon representing its motion over an interval of 100 million years. A program of exploration of the Moon with the aim of creating a permanent base on it is outlined. Such a base is intended for exploring the Earth, the Sun, and outer space.

Keywords: Moon, orbit, evolution, exploration, life, investigation, Earth, Sun, space

1. Introduction

The Moon is a satellite of the Earth; therefore, it is the body closest to our planet. After the Sun, the Moon exerts the second greatest impact on the Earth. For these reasons, the Moon will occupy the first place in the future exploration of space.

Since the Moon always faces the Earth from one side, its main motion is orbital. Therefore, all the features of this motion are of interest, including its evolution over long time intervals. The first part of this chapter is devoted to this problem.

In the second part, prospects for further space research are considered. More than half a century of experience in this research has shown that the effectiveness of studies strongly depends on the resource base invoked for performing the studies. Of the celestial bodies, the Moon is the body most suitable for creating a base on it. The chapter discusses a wide range of issues related to the feasibility of creating such a base, its structure, functioning, and prospects for research on it.

2. Evolution of the Moon's orbital motion

2.1 Coordinate system and orbital parameters

When solving the problem of the evolution of the rotational motion of the Earth over millions of years [1], it is necessary to have data on the coordinates of bodies acting on the Earth at any time from this time interval. The Moon exerts two-thirds of the total influence exerted on the Earth's rotational motion. The evolution of the Moon's orbital motion is therefore an important component of the posed problem.

The evolution of the orbits of Solar-system bodies can be determined by solving the problem of interaction for those bodies. For solving this problem, a Galactica system was developed [2–4]. The accuracy ensured by this software is several orders of magnitude higher than the accuracy ensured by other similar systems [5, 6]; this made it possible to solve the problem of the evolution of the Solar-system over 100 million years [7]. This problem is solved in the barycentric coordinate system *xyz* (**Figure 1**) attached to the fixed equatorial plane A_0A_0'. The origin *O* is located at the center of mass of the Solar-system. The results of solving this problem were saved in files following each 10 thousand years. Then, for those epochs, the interaction problem was solved, with the help of the Galactica program, per one orbital revolution of the body, and 11 parameters of its orbit were determined from its coordinates. For the Moon, the parameters of its orbit relative to the Earth are to be determined.

The Moon's orbital period is very short compared to that of planets. Therefore, the oscillation periods of the parameters of the Moon's orbit repeated many times over an interval of 10 thousand years. Therefore, with this interval, the evolution of the Moon's orbit was studied during its 736 continuous revolutions around the Earth, which took place during 56.7 years.

The position of the Moon's orbital plane, $\gamma_{Mo}A_1B$, is specified by its angle of inclination i_{Mo} to the plane of the stationary equator A_0A_0' and by the angle of the ascending angle $\varphi_{\Omega Mo}$, both defined in the caption to **Figure 1**. The position of the perigee of the Moon's orbit is specified by the angle φ_p. When analyzing the Moon's orbit, the origin *O* is located at the center of the Earth.

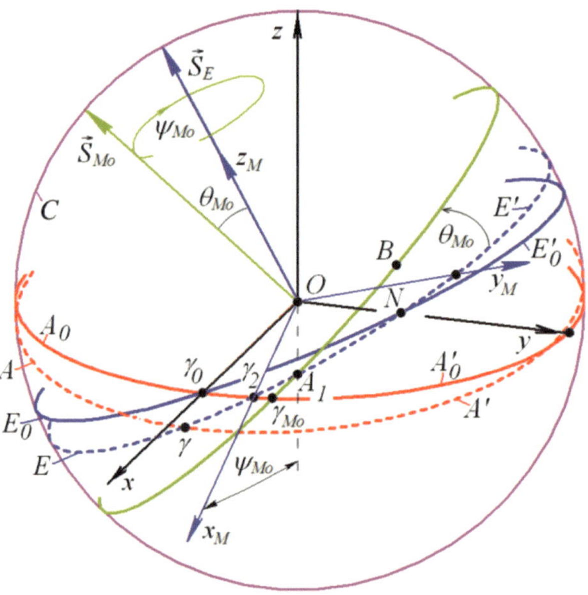

Figure 1.
The coordinate system and the main characteristics of the Moon's orbit: C *is the celestial sphere;* A_0A_0' *and* E_0E_0' *are the fixed planes of the equator and the Earth's orbit (ecliptic) for the epoch of 2000.0,* $JD_S = 2451545$; AA' *and* EE' *are the moving planes of the Earth's equator and orbit as of the current date;* $\gamma_{Mo}A_1B$ *is the Moon's orbital plane;* $\vec{S}_E$ *and* $\vec{S}_{Mo}$ *are the axes of the Earth's and Moon's moving orbits, respectively* EE' *and* $\gamma_{Mo}A_1B$, *which are perpendiculars to the orbital planes. The angles of the Earth's orbital plane relative to its equatorial plane* A_0A_0': $\varphi_{\Omega E} = \gamma_0\gamma_2$, $i_E = A_1\gamma_2\gamma_{Mo}$; *and same angles of the Moon's orbital plane relative to the Earth's equatorial plane:* $\varphi_{\Omega Mo} = \gamma_0\gamma_{Mo}$, $i_{Mo} = A_1\gamma_{Mo}A_0'$. B *is the projection of the perigee of the Moon's orbit on the celestial sphere, and* $\varphi_p = \gamma_{Mo}B$ *is its angular position;* ψ_{Mo} *u* θ_{Mo} *are the precession and inclination angles of the Moon's orbit relative to the moving plane of the Earth's orbit* EE'.

2.2 Dynamics of the Moon's orbit in the initial epoch

In the initial epoch $T = 0$, on the date of December 30, 1949 with the Julian-day number $JD_0 = 2433280.5$, consider the variation of the Moon's orbital parameters on a doubled interval of $\pm$ 736 revolutions, or $\pm$ 56.7 years. In order to distinguish between fluctuations, the results in **Figure 2** are shown for an interval of ± 10 years. The perigee radius R_p oscillates with a period T_{Rp} =0.5637 years around its average value $R_{pm} = 3.622069 \cdot 10^5$ km. The eccentricity of the orbit e oscillates with the same period around the mean value $e_m = 0.0563331$. In addition, there is a longer period of 3.719 years, yet exhibiting smaller oscillation amplitude.

The period of revolution of the Moon around the Earth P with respect to fixed stars, i.e. the sidereal period, oscillates around the average value $P_m = 7.47928 \cdot 10^{-2}$ years. There are two oscillation periods lasting 0.664039 and 3.719 years.

Over the entire interval, the perigee angle φ_p almost linearly increases into the future i.e. the perigee of the Moon's orbit rotates counterclockwise. The sidereal period of this rotation is $T_{\varphi p} = 8.8528$ years. In addition, the perihelion angle

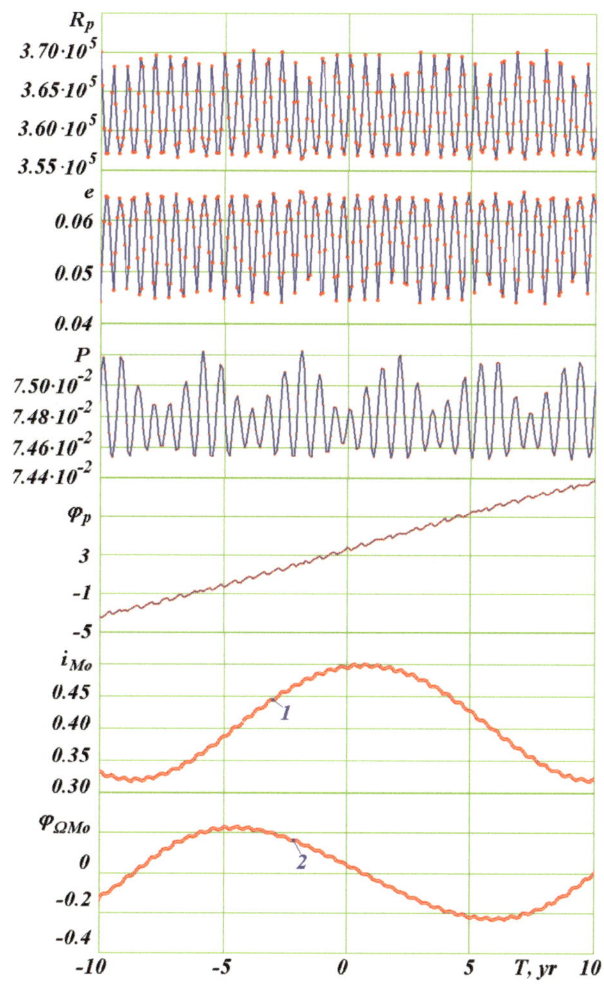

Figure 2.
Dynamics of the Moon's orbital elements in the geocentric equatorial coordinate system: perigee radius R_p *– in km, period* P *and time* T *– in sidereal years with a duration of 365.25636042 days, angles* φ_p, i_{Mo} *and* φ_{Mo} *– in radians; the white centerlines 1 and 2 are the approximating dependences (13) and (16), respectively. For other designations, see* ***Figure 1.***

oscillates with a short period $T_{\varphi p1}$ = 0.5637 years and with a long period $T_{\varphi p2}$ =18.6006 years.

The inclination angle i_{Mo} oscillates around its mean value i_{Mom} = 0.41526. The oscillations occur with two periods: a short one, of 0.4745 year, and a long one, of 18.6006 years. The angle of the ascending node $\varphi_{\Omega Mo}$ oscillates around the mean value of $\varphi_{\Omega Mom} = 6.5472 \cdot 10^{-4}$ with the same periods.

2.3 Precession of the Moon's orbital axis

When studying the orbits of the planets, we introduced the orbital axis $\vec{S}$ in the form of a unit-length vector normal to the orbital plane [8]. Using the inclination angle i_{Mo} and the ascending-node angle $\varphi_{\Omega Mo}$, we can write the projections of the orbital axis onto the axes of the xyz-coordinate system:

$$S_{Moz} = \cos i_{Mo}; \quad S_{Moy} = \sqrt{1 - S_{Moz}{}^2} \cos \varphi_{\Omega Mo}; \quad S_{Mox} = -S_{Moy} \mathrm{tg} \varphi_{\Omega Mo} \qquad (1)$$

The orbital axes of all planets precess about the angular momentum of all Solar-system bodies. As a result of the study, it was found that the axis $\vec{S}_{Mo}$ of the Moon's orbit precesses about the moving axis $\vec{S}_E$ of the Earth's orbit (**Figure 1**). The same will also be shown below. We introduce a coordinate system $x_M y_M z_M$. Along the z_M-axis of this system, the axis $\vec{S}_E$ is directed, and the axis x_M passes through the ascending node γ_2 of the Earth's orbit. Then, using the angles i_E and $\varphi_{\Omega E}$ specifying the position of the Earth's orbital plane and the projections of the Moon's orbital axis according to formulas (29) in Melnikov and Smulsky [7], one can find the projections S_{MoxM}, S_{MoyM}, and S_{MozM} of the axis $\vec{S}_{Mo}$ onto the axes of the $x_M y_M z_M$-coordinate system. **Figure 3a** shows the motion of the endpoint of the orbital axis $\vec{S}_{Mo}$as projected onto the $y_M x_M$-plane over the examined time interval of 113.4 years. It is evident from the graph that the endpoint of the vector $\vec{S}_{Mo}$ moves in a circle with slight fluctuations. The rotation period is T_S = −18.6006 years, and the oscillation period is $T_{\psi 1}$ = 0.4745 years. During the time interval under consideration, the axis $\vec{S}_{Mo}$ makes six revolutions in the clockwise direction.

From the projection onto the $z_M x_M$-plane (**Figure 3b**), it can be seen that the endpoint of the vector $\vec{S}_{Mo}$ executes small-amplitude oscillations along the z_M-axis with a swing of $\Delta z_M = 4.43 \cdot 10^{-4}$. Those oscillations are symmetrical about the x_M-axis.

Thus, the Moon's orbital axis precesses in a clockwise direction relative to the Earth's orbital axis. The precession period T_S is 18.6006 years. The precession proceeds with oscillations, which are called nutational. The period of the latter oscillations is $T_{\psi 1}$ = 0.4745 years.

Such studies were carried out for each epoch following 10 thousand years over time intervals of 0 ÷ −2 million years and −98 ÷ −100 million years. As a result, it was found that in all these cases the Moon's orbital axis $\vec{S}_{Mo}$ precesses relative to the moving axis $\vec{S}_E$ of the Earth's orbit (**Figure 1**).

In the $x_M y_M z_M$-coordinate system, the Moon's orbital axis $\vec{S}_{Mo}$ (**Figure 1**) is specified by the inclination and precession angles (respectively, θ_{Mo} and ψ_{Mo}):

$$\theta_{Mo} = \arccos S_{MozM}; \; ... \; \psi_{Mo} = \mathrm{arctg}\, S_{MoyM}/S_{MoxM} + 0.5\pi \qquad (2)$$

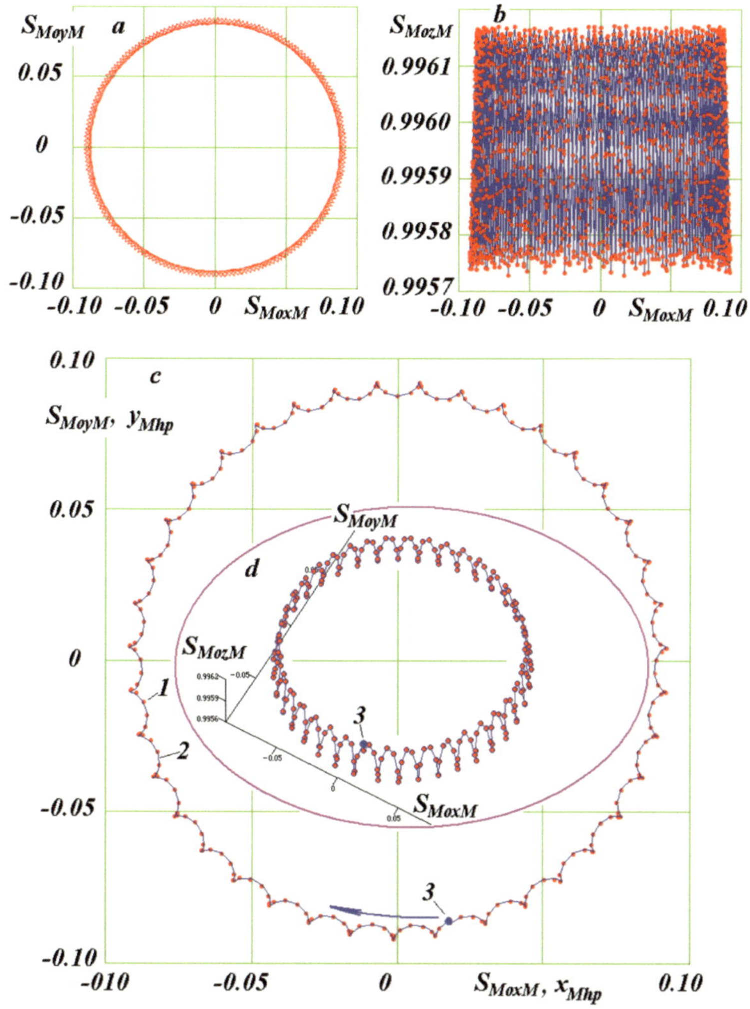

Figure 3.

Projections of the Moon's precession axis $\vec{S}_{Mo}$ onto the axes of the $x_My_Mz_M$ coordinate system over a period of 113.4 years (a, b) and from T = 0 to T = 18.6 years (c, d): d – in three-dimensional form; points: (1) projection of the axis $\vec{S}_{Mo}$ onto the y_Mx_M-plane; (2) hypocycloid equation $y_{Mhp}(x_{Mhp})$ (4)–(5); and (3) position of the axis $\vec{S}_{Mo}$ at T = 0.

Since the precession angle ψ_{Mo} varies over ranges wider than 2π, for calculating continuous values of this angle, its values at adjacent time intervals were calculated, and then the values obtained were summed using certain rules. As a result of studying the variation of the angle ψ_{Mo}, it was found that this angle decreases into the future, i.e., the axis $\vec{S}_{Mo}$ rotates in the clockwise direction with the rotation period of $T_S = -18.6006$ years. In addition, the angle ψ_{Mo} oscillates with a period of $T_{\psi 1} = 0.4745$ years and an amplitude of $\Delta\psi_{MoA} = 0.023662$. The inclination angle also oscillates with the latter period and with an amplitude $\theta_{MoA} = 0.002464 = 8.4692'$ about the mean value $\theta_{Mom} = 0.09006 = 5.1544°$.

Figure 3c shows, on a larger scale, the projection of the precessing orbital axis $\vec{S}_{Mo}$ onto the y_Mx_M-plane for one precession period $T_S = -18.6006$ years. Starting from the moment $T = 0$, marked by point 3, the end of the axis moves clockwise. It is evident from the graph that the endpoint of the vector $\vec{S}_{Mo}$ moves exactly around a circle, and the nutational oscillations here are regular. In **Figure 3d** dots and a line show the precession of the orbital axis in three dimensions. On the graph, the scale along the vertical z_M axis is increased 40 times.

As a result of an analysis, it was found that the endpoint of the vector $\vec{S}_{Mo}$ moves exactly along a hypocycloid. The hypocycloid is formed by some point of a circle of radius r rolling without slippage on the inner side of another circle of radius R.

In the $y_M x_M$-plane (**Figure 3c**), the radius of the great circle $R = \sin\theta_{Mom}$ is the mean value of the projection of the orbital axis $\vec{S}_{Mo}$ onto this plane, and the radius of the small circle $r = \sin\theta_{MoA}$ is the oscillation amplitude of this axis. The center of the small circle moves in the clockwise direction with the angular velocity $2\pi/T_S$. In this translational motion, the oscillations of the vector $\vec{S}_{Mo}$ occur with the period $T_{\psi 1}$, or with the angular velocity $2\pi/T_{\psi 1}$. Then, the absolute angular velocity of rotation of the small circle, $2\pi/T_n$, will be equal to the sum of these velocities: $2\pi/T_n = 2\pi/T_{\psi 1} + 2\pi/T_S$. That is why the period of the nutational rotation will be

$$T_n = \frac{T_S \cdot T_{\psi 1}}{T_S + T_{\psi 1}} = 0.48692 \text{ years} \tag{3}$$

Then, in the $y_M x_M$-plane the equation of the hypocycloid can be written as follows:

$$x_{Mhp} = R\cos\left(\varphi_{10} + 2\pi\frac{T}{T_S}\right) + r\cos\left(\varphi_{20} + 2\pi\frac{T}{T_n}\right) \tag{4}$$

$$y_{Mhp} = R\sin\left(\varphi_{10} + 2\pi\frac{T}{T_S}\right) + r\sin\left(\varphi_{20} + 2\pi\frac{T}{T_n}\right) \tag{5}$$

where $\varphi_{10} = 4.92766$ and $\varphi_{20} = 2.19315$ are the initial phases that specify the position of the vector $\vec{S}_{Mo}$ on the circles at the initial time $T = 0$.

The line in **Figure 3c** shows the trajectory of the motion along the hypocycloid, given by Eqs. (4) and (5), and the points are the projection of the motion of the Moon's orbital axis. Both are perfectly coincident. Thus, the Moon's orbital axis $\vec{S}_{Mo}$ executes an averaged clockwise motion around the Earth's orbital axis $\vec{S}_E$ with a period $T_S = -18.6006$ years. Here, the average angle between the axes $\vec{S}_{Mo}$ and $\vec{S}_E$ is $\theta_{Mom} = 5.1544°$.

The orbital axis $\vec{S}_{Mo}$ executes a second counterclockwise rotational motion about the averaged motion with a period T_n and an angular deviation from the median axis $\theta_{MoA} = 8.4692'$. During the complete revolution of the averaged axis of the Moon's orbit, $-T_S/T_{\psi 1} = 39.2$ nutational revolutions of the instantaneous axis $\vec{S}_{Mo}$ occur.

The dynamics of the inclination and precession angles, θ_{Mo} and ψ_{Mo}, of the Moon's orbit over an interval of 20 years is shown in **Figure 4**. The oscillations of the angle θ_{Mo} are more regular than those of the angle i_{Mo} of inclination of the Moon's orbit to the equatorial plane. They are harmonic with one and the same oscillation period. The precession angle ψ_{Mo} executes similar oscillations. At the same time, it monotonically decreases, this decrease being indicative of a clockwise precession of the orbital axis $\vec{S}_{Mo}$. The light line shows the approximating time dependence of the precession angle

$$\psi_{Mo}(T) = \psi_{Mo0} + 2\pi \cdot T/T_S + \Delta\psi_0 \tag{6}$$

where $\psi_{Mo0} = 0.202798$, and $T_S = -0.186006$ is the period of precession of the Moon's orbital axis $\vec{S}_{Mo}$.

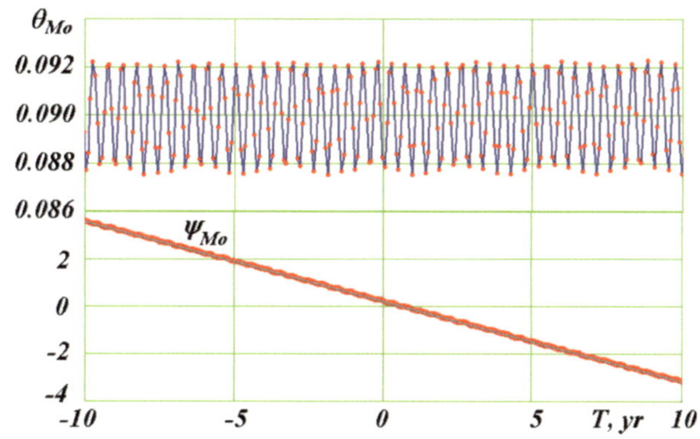

Figure 4.
Dynamics of the inclination and precession angles, θ_{Mo} and ψ_{Mo} of the Moon's orbit relative to the Earth's moving orbit. In the graph of ψ_{Mo}, the line shows the approximating dependence (7).

Changes in the Moon's orbit occur in the form of two groups of motions. In the first group, changes occur in the orbital plane with variation of the following parameters: perigee radius $R_p(T)$, perigee angle $\varphi_p(T)$, orbital eccentricity $e(T)$, and the orbital period $P(T)$. Here, the character T denotes the time dependence of the elements. In the second group, changes occur of the Moon's orbital plane $\gamma_{Mo}A_1B$ (**Figure 1**), specified by the angles $\varphi_{\Omega E}$ and i_{Mo} relative to the equatorial plane A_0A_0' or by the angles ψ_{Mo} and θ_{Mo} relative to the moving plane of the Earth's orbit EE'. Since the latter angles change more regularly, using them is more preferable in describing the motion of the Moon's orbital plane.

2.4 Approximation of the orbital-plane elements

As it was noted above, the behavior exhibited by the Moon's orbital elements was studied for the period of 736 its continuous revolutions in different epochs over the interval from 0 to 100 million years. In addition, the elements of the Moon's orbit were investigated following the adoption of different initial conditions in the integration of the equations of motion using the Galactica program [7]. As a result of these studies, regularities of the dynamics of the elements were established, and the approximating dependences for them were chosen. The final form of the approximations was refined on a doubled interval from −736 to +736 revolutions, in which the meantime falls onto the epoch of December 30.0, 1949 with the Julian-day number JD_0 = 2433280.5. The perigee radius is defined by the expression

$$R_p(T) = R_{pm} + R_{pA} \cdot \sin\left(\varphi_{Rp0} + 2\pi \cdot T/T_{Rp}\right) \quad (7)$$

where R_{pm} = 3.622069 · 10^5 km is the average value of the perigee radius, R_{pA} = 6.2754 · 10^5 km is the amplitude of oscillations, φ_{Rp0} = 0.942478 is the initial phase of oscillations, and T_{Rp} = 0.005637 is the period of oscillations of the perigee radius. The time T and the periods of oscillations are expressed here in sidereal centuries of 36525.636042 days per century, and they are counted from the JD_0 epoch, December 30, 1949.

As it is evident from the graph $R_p(T)$ in **Figure 2**, there are oscillation beats of the perigee radius, which can be described by invoking a second harmonic with a large period. However, due to the irregularity of these beats over large time intervals, the second harmonic did not significantly improve the approximation of the perigee radius.

The eccentricity is approximated with two harmonics:

$$e(T) = e_0 + e_{A1} \cdot \sin(\varphi_{e01} + 2\pi \cdot T/T_{e1}) + e_{A2} \cdot \sin(\varphi_{e02} + 2\pi \cdot T/T_{e2}) \tag{8}$$

whose characteristics are given in **Table 1**.

The perigee of the Moon's orbit rotates counterclockwise and, in addition, it executes oscillatory movements, which were also approximated with two harmonics:

$$\varphi_p(T) = \varphi_{p0} + 2\pi \cdot T/T_{\varphi p} + \Delta\varphi_{p01} + \varphi_{pA1} \cdot \sin\left(\varphi_{p01} + 2\pi \cdot T/T_{\varphi p1}\right) + \Delta\varphi_{p02} + \varphi_{pA2} \cdot \sin\left(\varphi_{p02} + 2\pi \cdot T/T_{\varphi p2}\right) \tag{9}$$

where $T_{\varphi p}$ is the period of revolution of the Moon's perigee, and $T_{\varphi p1}$ and $T_{\varphi p2}$ are the first and second periods of oscillations of the perigee angle. The coefficients entering Eq. (9) are given in **Table 2**.

The Moon's orbital period P oscillates around its mean value P_m. The analysis of these oscillations was carried out considering the relative difference $\delta P = (P - P_m)/P_m$. Since the period P and the semi-major axis a of the Moon's orbit vary consistently, the analysis of those oscillations in relative differences allows their consistent approximation. The period P is also approximated with two harmonics:

$$P(T) = P_m\left[\left(1 + \Delta P_0 + \Delta P_{A1} \cdot \sin\left(\varphi_{pr01} + 2\pi \cdot T/T_{p1}\right) + \Delta P_{A2} \cdot \sin\left(\varphi_{pr02} + 2\pi \cdot T/T_{p2}\right)\right]\tag{10}$$

where T_{p1} and T_{p2} are the first and second oscillation periods of the orbital period, and the values of the coefficients are given in **Table 3**.

e_0	e_{A1}	φ_{e01}	T_{e1}
0.0563331	0.0113634	−2.19911	0.005637
	e_{A2}	φ_{e02}	T_{e2}
	6.91384 E−4	−1.5708	0.03719

Table 1.
Coefficients in Eq. (8).

φ_{p0}	$T_{\varphi p}$	$\Delta\varphi_{p01}$	φ_{pA1}	φ_{p01}	$T_{\varphi p1}$
3.67159	0.088528	1.34024E−4	0.200529	2.19911	0.005637
		$\Delta\varphi_{p02}$	φ_{pA2}	φ_{p02}	$T_{\varphi p2}$
		−4.91312E−3	0.196967	−0.188496	0.186006

Table 2.
Coefficients in Eq. (9).

P_m	ΔP_0	ΔP_{A1}	φ_{pr01}	T_{p1}
7.479277E−4	1.01403E−4	0.00385003	0.628319	0.00664039
		ΔP_{A2}	φ_{pr02}	T_{p2}
		0.00141509	−1.41372	0.03719

Table 3.
Coefficients in Eq. (10).

Evidently, some parameters have identical oscillation periods. Perigee radius $R_p(T)$, eccentricity $e(T)$, and perigee angle $\varphi_p(T)$ have identical first periods of 0.005637 century, whereas the eccentricity and the orbital period $P(T)$ have identical second oscillation period of 0.03719 century.

2.5 Approximation of the orbital angles

As a result of studies, it was found that the precession angle ψ_{Mo} oscillates with two periods, a shorter (0.4745 years) and a longer (2.995 years) one. Since the amplitude of the large-period oscillations is small, we neglect those oscillations. As a result, the precession angle can be approximated with the following expression:

$$\psi_{Mo}(T) = \psi_{Mo0} + 2\pi \cdot T/T_S + \Delta\psi_{Mo0} + \Delta\psi_{MoA} \cdot \sin\left(\varphi_{\psi} + 2\pi \cdot T/T_{\psi 1}\right) \quad (11)$$

where ψ_{Mo0} = 0.202798, T_S = −0.186006 is the precession period of the Moon's orbital axis $\vec{S}_{Mo}$, $\Delta\psi_{Mo0}$ = 2.3024710^{-4}, $\Delta\psi_{MoA}$ = 0.023662, φ_{ψ} = 2.82743, and $T_{\psi 1}$ = 0.004745 is the period of oscillations of the precession angle ψ_{Mo}.

The inclination angle θ_{Mo} also oscillates with two periods. The longer period, equal to 2.995 years, has an amplitude of $5.978 \cdot 10^{-5}$ radians, which value is almost two orders of magnitude smaller than the amplitude of the first period. Therefore, the second harmonic, i.e. the one with the period of 2.995 years, can also be neglected, and the approximation for the nutation angle, therefore, has the form:

$$\theta_{Mo}(T) = \theta_{Mo0} + \theta_{MoA} \cdot \sin\left(\varphi_{\theta} + 2\pi \cdot T/T_{\psi 1}\right) \quad (12)$$

where θ_{Mo0} = 0.09006, θ_{MoA} = 0.002464, and φ_{θ} = −2.19911

The angles ψ_{Mo} and θ_{Mo} are tied to the moving plane of the Earth's orbit EE' (see **Figure 1**), so they are inconvenient to use. We, therefore, pass to the angles $\varphi_{\Omega Mo}$ and i_{Mo}, which specify the position of the Moon's orbital plane relative to the fixed plane of the equator A_0A_0' (**Figure 1**), with which the main coordinate system xyz is associated. In the spherical triangle $\gamma_2\gamma_{Mo}A_1$, the side γ_2A_1 = ψ_{Mo} and the two angles γ_2 = i_E and A_1 = θ_{Mo} are known. The cosine theorem can be used to determine the obtuse angle $\gamma_2\gamma_{Mo}A_1$, from which the acute angle i_{Mo} can be subsequently found: i_{Mo} = $\pi - \gamma_2\gamma_{Mo}A_1$. As a result, for the angle of inclination of the Moon's orbital plane to the plane of the stationary equator, we obtain the following expression:

$$i_{Mo} = \pi\text{–arccos}\left(-\cos\, i_E \cdot \cos\theta_{Mo} + \sin\, i_E \cdot \sin\theta_{Mo} \cdot \cos\psi_{Mo}\right) \quad (13)$$

As it is seen from **Figure 1**, the angle specifying the position of the ascending node of the Moon's orbit is equal to the sum of two arcs,

$$\varphi_{\Omega Mo} = \gamma_0\gamma_{Mo} = \varphi_{\Omega E} + \gamma_2\gamma_{Mo} \quad (14)$$

By the sine theorem, in the triangle $\gamma_2\gamma_{Mo}A_1$ we have

$$\sin\,\gamma_2\gamma_{Mo}/\sin\theta_{Mo} = \sin\psi_{Mo}/\sin\left(\pi\text{–}i_{Mo}\right) \quad (15)$$

and, therefore, the arc $\gamma_2\gamma_{Mo}$ can be found. Then, according to Eq. (14), the position of the ascending node can be found as

$$\varphi_{\Omega Mo} = \varphi_{\Omega E} + \text{arcsin}\left[\sin\psi_{Mo} \cdot \sin\theta_{Mo}/\sin\left(\pi\text{–}i_{Mo}\right)\right] \quad (16)$$

In order to check the validity of the obtained approximations of the Moon's orbital elements (13) and (16), we superimposed onto **Figure 2** the calculated

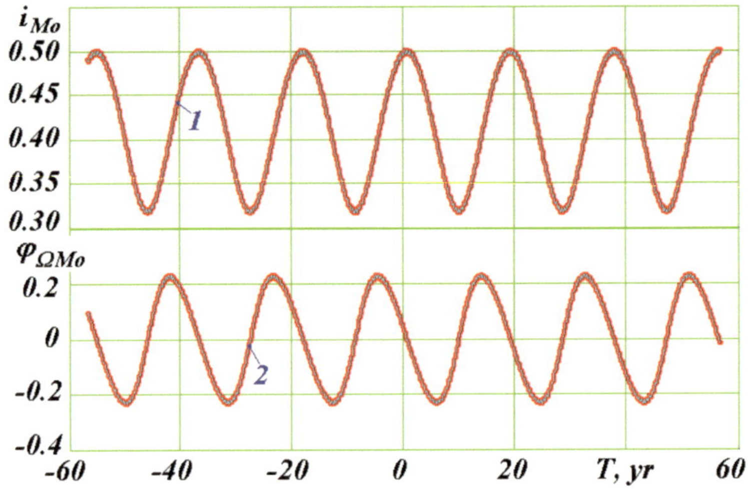

Figure 5.
Comparison of the dynamics of the angles i_{Mo} *and* $\varphi_{\Omega Mo}$*, specifying the position of the Moon's orbital plane relative to the equatorial plane, as obtained in two ways: thick lines – numerical integration; light thin lines 1 and 2 – approximating dependences (13) and (16), respectively.*

elements that were obtained using the Galactica program for the integration of the equations of motion. **Figure 5** shows, over the entire interval of $\pm$ 56.7 years, the dynamics of the angles i_{Mo} and $\varphi_{\Omega Mo}$ obtained by two methods: using numerical integration (thick lines) and using approximations (13) and (16) (light line). As it is seen from the graphs, the approximations yield data perfectly coincident with the short- and long-period oscillations of the angles i_{Mo} and $\varphi_{\Omega Mo}$. Thus, this check has fully confirmed the validity of the adopted approximations.

2.6 Evolution of orbital elements over an interval of 100 million years

So, the dynamics of Moon's orbital elements R_p, e, φ_p, P, i_{Mo}, and $\varphi_{\Omega Mo}$ relative to the fixed plane of the equator in the geocentric coordinate system xyz is described by Eqs. (7)–(10), (13), and (16). This description was obtained over a time interval of 113.4 years. As already mentioned, for establishing the validity of this description over large time intervals, studies were carried out over intervals of $0 \div -2$ million years and $-98 \div -100$ million years. Following each 10 thousand years, the dynamics of the Moon's orbital elements were investigated during 736 continuous orbital revolutions of the Moon. The dynamics in different epochs did not differ qualitatively from that shown in **Figure 2**. With the purpose of comparison of those dynamics, the average values of individual elements during 736 orbital revolutions were calculated. Then, the evolution of these average values, as well as the periods of rotation, periods of oscillations, and oscillation amplitudes overtime periods of 2 million years with an interval between points of 10 thousand years, was investigated.

As an example, **Figure 6** shows the evolution of the average orbital period P_m, eccentricity e_m, inclination angle θ_{Mom}, and the amplitude θ_{MoA} of nutational oscillations. The graphs show the relative changes of these quantities. These changes were determined in the same way, for example, for the average period of the Moon's orbital revolution this value was calculated as follows:

$$\delta P_m = (P_m - P_{m0})/P_{m0} \tag{17}$$

where P_{m0} is the value of the average orbital period over 736 orbital revolutions of the Moon in the modern epoch. In calculating the relative changes of the

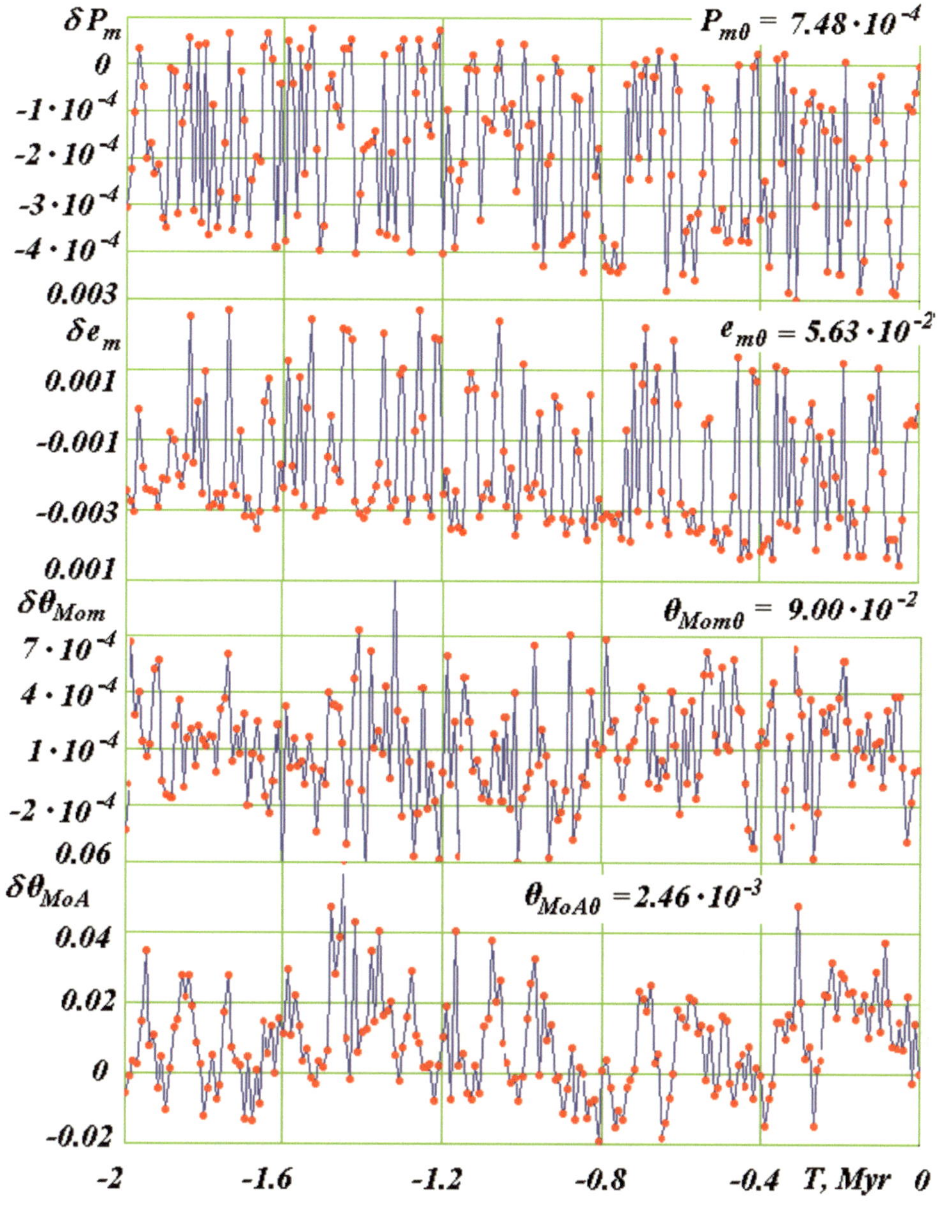

Figure 6.
Evolution, over the period of 2 million years, of relative averages for 736 revolutions of the deviations of Moon's orbital parameters: period δP_m*, eccentricity* δe_m*, inclination angle* $\delta\theta_{Mom}$*, and the amplitude of nutational oscillations* $\delta\theta_{MoA}$*;* T *– in million years.*

amplitudes ($\delta\theta_{MoA}$), instead of the mean values entering Eq. (17), the amplitude θ_{MoA} yielded by approximation (12) was used.

As it is seen from **Figure 6**, the oscillation amplitude of the relative mean δP_m, δe_m and $\delta\theta_{Mom}$ are $2{\cdot}10^{-4}$, 0.003, and $4.5{\cdot}10^{-4}$, respectively. At the same time, the similar relative oscillation amplitudes during 736 Moon's orbital revolutions are $3.85{\cdot}10^{-3}$, 0.2, and $2.7{\cdot}10^{-2}$, respectively. Thus, the analyzed fluctuations of Moon's parameters P, e, and θ_{Mo} exceed their changes over the interval of $0 \div -2$ million years by factors of 19, 67, and 60, respectively. This conclusion is also confirmed by the graph $\delta\theta_{MoA}(T)$ in **Figure 6**: over the interval of $0 \div -2$ million years, the amplitude of nutational oscillations θ_{MoA} fluctuates within 2%.

The rest approximation parameters exhibit similar behavior. Similar results were obtained for the interval of −98 ÷ −100 million years. This allows us to conclude that, over the interval of 0 ÷ −100 million years, if there occur oscillations with longer periods than those used in our approximations, then the amplitude of such oscillations does not exceed a few percent of the considered oscillation amplitudes.

2.7 Mathematical model for the Moon's motion

Thus, Eqs. (7)–(10), (13), (16) describe the evolution of Moon's orbital elements R_p, e, φ_p, P, i_{Mo}, and $\varphi_{\Omega Mo}$ which can be used over the interval 0 ÷ −100 million years. We have developed a mathematical model of body motion in an elliptical orbit [9], which is based on the listed orbital elements. That is why this model, with Eqs. (7)–(10), (13), and (16), allows one to calculate the Moon's coordinates in the equatorial coordinate system at any time in the interval of 0 ÷ −100 million years.

Figure 7 compares the Moon's orbits calculated using this model with a time step of $1{\cdot}10^{-4}$ years and numerical integration performed with the help of the Galactica program. The same orbital comparisons were made for the planets [9]. The orbits of the planets calculated by the mathematical model are no visual difference from the orbits obtained by numerical integration. As it is seen from **Figure 7**, such differences are observed for the Moon's orbit. This is due to the shorter Moon's orbital period compared to that of the planets. Nevertheless, this mathematical model of the Moon made it possible to solve the problem of the evolution of the Earth's rotational axis with acceptable accuracy. Comparison of the results of this problem for 200 thousand years, solved with this model of the Moon's orbit and without it, proved differences to be insignificant [1].

2.8 Comparison of calculations with observation data

The orbital periods of the Moon, the precession of its orbital axis, and the rotation of the perihelion oscillate about the average values of these quantities. Over

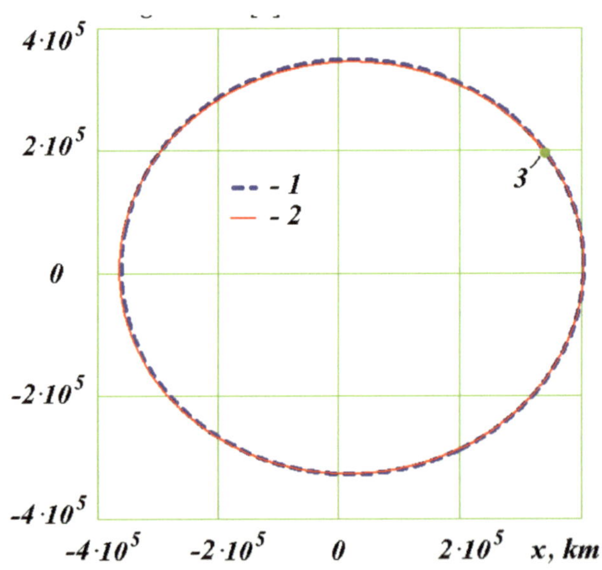

Figure 7.
Comparison of the projections of the Moon's orbit onto the equatorial plane xy *calculated in two ways: (1) based on the results of numerical integration, by the Galactica program, of the differential equations of motion of Solar-system bodies; (2) according to the mathematical model of the Moon's motion; and (3) the starting point of the orbit at the moment* T = 0.

the interval of 113.4 years, the average values were designated as P_m, T_S, and $T_{\varphi p}$, respectively. Their magnitudes in sidereal years are given in **Table 4**. Astronomy considers different months with durations expressed in days. The sidereal month with a period P_{msid} is specified relative to fixed stars. The synodic month with a period P_{msyn} is specified in relation to the Earth. The sidereal orbital period of the Earth relative to stars is P_{Esid} = 365.25636042 days. Therefore, the angular velocity of the Moon in its orbit around the Earth relative to it is equal to the difference between the angular orbital velocities of the Moon and the Earth in relation to stars. Therefore, the duration of the synodic month is

$$P_{msyn} = \frac{P_{Esid} \cdot P_m}{P_{Esid} - P_m} \tag{18}$$

where the period $P_m = 7.479277 \cdot 10^{-2}$ sidereal years expressed in days is equal 27.318536.

The period P_{mano} of the anomalistic month is specified in relation to the Moon's perigee or its apogee. The period of motion of the Moon's perigee relative to stars is denoted as $T_{\varphi p}$. Therefore, the period of the anomalous month is

$$P_{mano} = \frac{T_{\varphi p} \cdot P_m}{T_{\varphi p} - P_m} \tag{19}$$

where the period $T_{\varphi p}$ is expressed in days.

The Draconic month with a period P_{mdra} is specified in relation to the Moon's ascending node. The position of the ascending node γ_{Mo} is specified by the angle $\varphi_{\Omega Mo}$ (**Figure 1**), and its motion relative to fixed stars occurs with the precession period T_S of the Moon's orbital axis $\vec{S}_{Mo}$. Therefore, the draconic-month period is

$$P_{mdra} = \frac{T_S \cdot P_m}{T_S - P_m} \tag{20}$$

where the period T_S is expressed in days.

A tropical month with a period P_{mtro} is defined in relation to the Earth's moving equator AA' in **Figure 1**. The moving equator, as well as the Earth's axis of rotation, precess relative to fixed stars with a period of P_{prEax} = −25738 sidereal years [1]. Therefore, the period of the tropical month will be:

Method	P_m	T_S	$T_{\varphi p}$	P_{msid}
		Siderial years		**Days**
Calculation	$7.479277 \cdot 10^{-2}$	−18.60062	8.852804	27.318536
Observation	—	—	—	27.321662
Relative difference, δ	—	—	—	$-1.14 \cdot 10^{-4}$
Method	P_{msyn}	P_{mano}	P_{mdra}	P_{mtro}
		Days		
Calculation	29.526938	27.551303	27.209129	27.315564
Observation	29.530589	27.554550	27.21221	27.321582
Relative difference, δ	$-1.24 \cdot 10^{-4}$	$-1.18 \cdot 10^{-4}$	$-1.13 \cdot 10^{-4}$	$-1.14 \cdot 10^{-4}$

Table 4.
Comparison of calculated and observed average durations of various months: sidereal P_{msid}, *synoptic* P_{msyn}, *anomalistic* P_{mano}, *and draconian* P_{mdra}.

$$P_{mtro} = \frac{P_{prEax} \cdot P_m}{P_{prEax} - P_m} \tag{21}$$

where the period P_{prEax} is expressed in days.

These periods, as calculated by Eqs. (18)–(21) and as evaluated from the observations of [10] are summarized in **Table 4**. The relative difference between the calculated and observed periods is expressed in terms of a parameter δ defined similarly to Eq. (17). As it is seen, the largest value of δ is $1.24 \cdot 10^{-4}$. The main contribution to this difference is made by the sidereal period P_m of Moon's orbital revolution. If we use the observed value of P_m = 27.321662 days, then the δ-values will decrease by two-five orders of magnitude.

As it is seen from **Figure 2**, the Moon's orbital period P experiences oscillations with relative amplitudes ΔP_{A1} and ΔP_{A2}, which in total make up 0.0053 part of the period P. In addition, from **Figure 6** it is seen that over time intervals of tens of thousands and more years there exist oscillations of the average period P_m with a relative amplitude of the order of $2 \cdot 10^{-4}$. For oscillating quantities, their average values depend on the interval over which the averaging is performed. The value of P_m given in **Table 4** was obtained by an averaging performed over an interval of 113.4 years, and the value of the sidereal period in astronomy has an averaging interval of about 2 thousand years. This seems to be the main reason for the difference between the calculated and observational data with a relative value of the order of $1 \cdot 10^{-4}$.

3. Prospects for on-Moon research

3.1 Problems, and their content and structure

There are various proposals for research to be carried out on the Moon. Some of those proposals may prove useful, while others, not [11]. The Moon near the Earth is the only body close to it. Therefore, not counting the Earth, the Moon is the only body that can be used for the study and exploration of outer space. It seems that such activities should be carried out along three lines. It is necessary to study the Earth, the Sun, and outer space from the Moon. For this purpose, an Earth Service should be established on the visible side of the Moon, and a Space Service, on its opposite side. Solar exploration will be additionally performed by both Services.

The mission of the Earth Service is to continuously monitor and analyze all processes and phenomena that occur on the Earth. Observations should be carried out using optical means in all ranges of the spectrum. In addition, other available methods known in astronomy for measuring the physical characteristics of the Earth, such as the methods of radio astronomy, γ-astronomy, methods for measuring the magnetic properties of the Earth's surface, and others, should be used. The results of such measurements will provide a better insight into the processes taking place on the Earth. As a result, it will become possible to improve methods for the long-term forecasting of weather and such catastrophic events as tropical cyclones, hurricanes, typhoons, etc. Continuous observations of the Earth will provide reliable data on many events and processes occurring on the planet: the state of ice conditions in the southern and northern oceans, the dynamics of snow cover, various seasonal changes of the Earth's surface, fire hazard of territories, volcanic eruptions, man-made accidents and disasters, the fall of large meteorites, as well as various military actions on a global scale.

All this will contribute to a safer and more stable habitation of humans on the Earth.

The Solar Service, located on both opposite hemispheres, will allow the observation of processes on the Sun in an almost continuous mode. Solar flares affect the dynamics of the Earth's atmosphere, and they presently cause many dangerous atmospheric phenomena [12]. The Sun's activity, manifested in the number of sunspots, varies periodically. Such periods correlate with the periods of the Sun's movement around the center of mass of the Solar-system [12, 13]. Their duration is 22 years with two sub-periods each lasting 11 years. In addition, there are also large periods lasting hundreds of years. Possibly, those fluctuations of the Sun's activity cause the short-period variations of the Earth's climate [13].

The study of solar processes will allow a more detailed understanding of the processes occurring on stars. The two Solar Services will host the equipment used for studying the Sun and stars from the Earth. The effectiveness of the use of this equipment on the Moon is expected to be much higher, as there is no cloudiness and no atmosphere there. Due to the small force of gravity, structures cumbersome on the Earth will appear weighing much less on the Moon.

The Space Service is the most important part of human activity on the Moon. The importance and relevance of its tasks to the solution of many challenging problems will permanently grow in time. At an early stage, this service will carry out all studies currently being carried out on the Earth with the help of Earth's satellites. As this service evolves, these tasks will be supplemented with new ones that cannot be accomplished with the help of satellites. One of such tasks is the communication with spacecraft sent into deep space. The absence of atmosphere and intrinsic magnetic field on the Moon will make it possible to carry out such connections in a more stable manner.

What divisions should be included in these two services? Each service should consist of the following three departments: (1) Research Department; (2) Engineering Department; and (3) Greenhouse Department.

The task of the Research Department is to carry out works on the study of the Earth, the Sun, and space. The task of the Engineering and Technical Departments is to create the material base of the service and ensure its functioning. The task of the Greenhouse Department is to support life on the Moon, provide food for inhabitants, and to ensure life in all structures of the greenhouse economy.

At the first stage, the tasks of the Greenhouse Department will come as the main ones, since the human civilization presently has no experience in supporting life in extraterrestrial conditions. Work on the Earth and artificial earth satellites to create life in artificial conditions should begin in advance. Some experience in this area already exists. It is necessary to study this experience and formulate a research program for the creation of various life elements in extra-terrestrial conditions in relation to the Moon. After accomplishing this work, we can initiate the development of a greenhouse project on the Moon.

Until the full-fledged functioning of the greenhouse economy begins on the Moon, research and engineering works will mainly be carried out with the help of automatic machines and mechanisms controlled from the Earth.

3.2 Transportation on the Moon

For moving on the Moon, it will be necessary to create walking and running vehicles. Animals on the Earth, two- and four-legged, can move at a decent speed comparable with the speed of wheeled vehicles. But an animal can move at this speed in off-road conditions. When moving, the animal observes its path and puts its foot on the ground taking into account all the circumstances arising at the point of contact with the ground. Modern means of observation, monitoring, and control make it possible to create a mechanical leg of a vehicle that will function no worse than the leg of the fastest animal. In further development, a vehicle with mechanical legs will reach in off-road conditions the speed of a wheeled vehicle on a good road.

Such vehicles with mechanical legs can be supplemented with mechanical arms or some legs can be provided with the function of arms. Mechanical arms will help the vehicle to extricate itself from emergency situations: when overturning, when driving in dangerous areas, etc. Control algorithms shall be developed for different situations and with time the reliability of such vehicles will approach 100%.

When driving on established routes, a vehicle with mechanical arms can clear the most disturbing obstacles out of the way. In this way, paths and roads for this transport will be created, along which the speed of movement will be increased.

Such vehicles, equipped with navigation aids, will be able to move with or without man. All works related to the delivery of goods will be executed without people. This will greatly simplify, and reduce the cost of, moving goods, since there will be no need in using life support systems for people.

Long-distance movements, for example, those between the Earth and the Space Services, will be performed using jet engines along ballistic trajectories. In jet engines on the Earth, fuel burns in an oxidizing agent, the combustion products acquire a high speed, and the jet stream propels the vehicle, for example, a spacecraft. In lunar jet engines, lunar sand and dust will be used as the jet substance. The jet vehicle must possess the energy required to impart the speed of the jet stream to this material. This energy can be the electrical energy stored in batteries. The batteries will be charged by solar panels during the lunar day.

The acceleration of the substance can be carried out electrically. For example, a charge of one sign can be imparted to a bulk material, which then enters an interelectrode space with a high voltage to undergo acceleration. In the mechanical method, the bulk material is fed to a rotating device to acquire the required speed. In this case, in order to prevent the vehicle from rotating, it is necessary to have paired devices rotating in different directions.

As the bulk material, lunar regolith can be used, which, apparently, includes terrestrial analogs in terms of its granulometric composition such as dust, powder, sand, and sandy loam.

The issue of obtaining and storing energy is a special problem that requires careful study. Apparently, in the non-polar regions of the Moon, solar energy will be sufficient. Solar panels can provide electricity that needs to be stored for the Moon night. For heating during the night and for cooling during the day, respectively heat and cold accumulators must be used. Electricity can also be generated based on the temperature difference between the lunar surface and the constant-temperature layer beneath it. This temperature difference exists both during the day and at night. Apparently, Stirling engines can be used here for doing work and for generating electricity.

3.3 Materials and substances

For the creation of the Earth and Space Services, various materials and substances are needed. Consider what is required for supporting life on the Moon. Greenhouses will need soil, water, and air to function. Soil samples can be taken from the Earth. When plants settle on them, the soil can be mixed with lunar soil, with its amount being gradually increased. Apparently, not every lunar soil is suitable for these purposes. Therefore, a lot of work needs to be done to study the lunar soil, prepare and collect the required composition, and deliver it to greenhouses.

Where can we get water? During lunar days, the Moon's surface gets heated, and the water boiled away and evaporated. It is necessary to study the distribution of temperature over the lunar surface. Somewhere closer to the poles, a negative temperature can be found. It might be possible to find ice there.

In equatorial and middle latitudes, the temperature of the lunar surface varies from hundreds of Celsius degrees during the day to hundreds of degrees below zero at night. But with depth, the layer of variable temperature must vanish, and a constant temperature must establish. How low is this temperature? If the temperature is negative, then there may be ice found at this depth.

Thus, in order to find water, one has to carry out temperature studies of the Moon, both in-depth and over the surface.

Where can we get air? On the Earth, air contains 80% of nitrogen and 20% of oxygen. There are also small amounts of other gases. Apparently, many of them are not necessary.

There are no ready air and component gases (nitrogen and oxygen) on the Moon. Therefore, they must be obtained from substances available on the Moon. It is necessary to study the composition of lunar rocks. Then, people on the Earth must develop technologies for the extraction of nitrogen and oxygen from these rocks. Subsequently, the composition of the artificial air can be optimized with the help of plants and algae. Among them are those that give off oxygen as well as other gases.

For the construction of a greenhouse, structural materials, metals, and various substances are needed. It is impossible to get them from the Earth. From the Earth, it will be necessary to transport finished products, complex instruments and tools, machines, and similar products, which are impossible to manufacture on the Moon. All necessary materials and substances must be extracted from minerals available on the Moon. That is why the Moon's geology must be well studied. On its basis, processes on the transformation of lunar minerals into necessary materials and substances should be developed on the Earth.

3.4 Safety of buildings on the Moon

Buildings on the Moon will require a lot of spent effort, money, and time. Therefore, they must be durable with a service life amounting to hundreds of years. In this regard, it will be necessary for people on the Moon to protect themselves from natural disasters. This can be soil creeps on slopes, rockfalls, meteorite falls, etc. Some of such processes and events can pose no real threat. That is why, before the start of construction, it will be necessary to perform a study of possible risks and their occurrence probabilities. As for the meteorite danger, its reality is beyond doubt, since the entire surface of the Moon, like that of all celestial bodies, is dotted with meteorite craters. Therefore, this threat must be treated with close attention. Apparently, it is necessary to conduct experimental observations on the probability, composition, and characteristics of meteorites falling onto the Moon. For this purpose, it is possible to spread a screen on the Moon's surface with means for observing and controlling the fall of meteorites. Information from such devices must be transmitted to Earth. Observations should be made over several years. They will allow scientists to obtain data on meteorite hazards, which is necessary for the design of buildings. There should be two such sites in the places of proposed construction: one on the visible side of the Moon, and the other on the opposite side.

Over the long service life of structures, there will always be a danger of being hit by large meteorites. Therefore, a vitally important part of the greenhouse must be created below the Moon's surface. Apparently, the best option would be the creation of each service near a rock hill. The greenhouse farm will be located outside the hill, with its all vitally important systems being hidden in hollow rooms inside the hill. The top of the hill will provide the reliable protection of such systems from relatively large meteorites. The greenhouse should be made sectioned. Then, in the event of a depressurization having occurred at some section as a result of a meteorite hit, the remaining sections will automatically be cut off from the damaged section and continue to function.

3.5 The relations between humans in their activities on the Moon

Services on the Moon will be created in the interests of all mankind. However, there are states on Earth the relations among which cannot be called friendly. Mutual threats are possible and wars to destroy each other are not ruled out. This situation may not radically change in the next hundreds of years. Therefore, principles to govern the relations between people of the Earth during their activities on the Moon must be formulated. Based on the conditions necessary for the successful functioning of two services on the Moon, let us try to formulate some of those principles.

First, each state has the right to take part in the creation and functioning of these services, and it will share the results obtained.

Secondly, since there are two services, it makes sense to form two groups of states, one being responsible for the service on the visible side of the Moon, and the other, on its backside.

Thirdly, having obtained permission, the representatives of one group of states will have the right to visit the territory of the service shared by the second group of states.

Fourth, each group of states shall share its achievements and results with the members of the other group at no cost.

Fifth, unfriendly and hostile relations among states on the Earth shall not be practiced by representatives of such states on the Moon.

Those who call to violate this principle will be subject to capital punishment with no statute of limitations.

Mankind already has experience of such cooperation gained in the study of Antarctica, in the Apollo-Soyuz project, and in the activities at the International Space Station. This experience can be considered successful. For cooperation on the Moon, the accumulated experience shall be widely applied.

3.6 Work sequence

Moon exploration began 50 years ago by the Soviet Union and the United States. Other countries now take part in it. This activity will be continued by different countries in the future. For making those fragmented studies fruitful, it is necessary to set common goals and formulate certain tasks. Then all the studies will add to our common knowledge of the Moon, which subsequently will allow these goals to be achieved.

Therefore, it is necessary to conduct an international discussion of the problem of Moon exploration by all interested parties. The result of this discussion should be the establishment of an International Committee for Moon Exploration. The first task of this committee shall be the development of a preliminary project on the prospects for Moon exploration.

In this project, all the goals and objectives discussed above will be concretized. This will allow different countries to unite their efforts. The International Committee will have the task of coordinating these studies, analyzing and summarizing their results, and setting further tasks.

This work will contribute to the rapprochement of the individual parties, uniting them in the implementation of large projects. This cooperation will further lead to the consolidation of collaboration teams necessary for the creation of Earth and Space services.

One of these preliminary tasks is the creation of a Moon satellite. The satellite is needed as an intermediate station for flights from Earth to the Moon and back. In addition, a satellite is needed to connect the Space Service with the Earth, and the Earth and Space services with each other.

Further development of the International Committee for Moon Exploration will turn it into main mankind's organization on the exploration of the Moon and lunar works.

3.7 Possible missions to be performed using Moon services

When mankind starts establishing services on the Moon, the task may be set to provide the Moon with a long-term satellite. Previously, we have performed trajectory calculations for transforming the Apophis and 1950DA asteroids into Earth's satellites [14]. The task here is to choose an asteroid suitable for making it a Moon satellite. Apparently, the orbit of such a satellite should be circular or having a small eccentricity and a semi-major axis about 5000 km long. That is, the spacing between such an asteroid and the Moon should be equal to the above distance. The satellite's orbit must lie in the Moon's orbital plane. Such a satellite will increase the reliability of movements between the Earth and the Moon.

In astronomy, various methods are used to determine the distance from the Earth to astronomical objects. The most reliable one is the triangulation method, in which the angles of observation of a star from opposite points in the Earth's orbit are measured. The angles can be determined from the displacement of a star over the celestial sphere against the background of more distant stars. In this way, one can measure the distance to objects located at a distance of 20 parsecs (pc). In this case, the base distance is the semi-axis of the Earth's orbit a. On increasing the base length, the range of measured distances will increase in proportion to this length.

One can increase the base by placing one of the observation points on a spacecraft launched from the Earth along a hyperbolic orbit. The location of the star, observed on the spacecraft at some distance r from the Earth and communicated to it, will make it possible to determine the distance to stars located at typical distances greater than 20 pc by a factor of r/a.

We assume that a spacecraft is launched at point A in **Figure 8** in the Earth's orbital plane in the direction of Earth's orbital motion. Suppose, for instance, that the speed of the spacecraft relative to the Earth is 20 km/sec, and its speed relative to the Sun is 50 km/sec. At this speed, the spacecraft moves in a hyperbolic orbit, with its speed at infinity being v_∞ = 28 km/s, i.e. the spacecraft leaves the Solar system at this speed. Six months later, a similar spacecraft is directed at point B in the opposite direction.

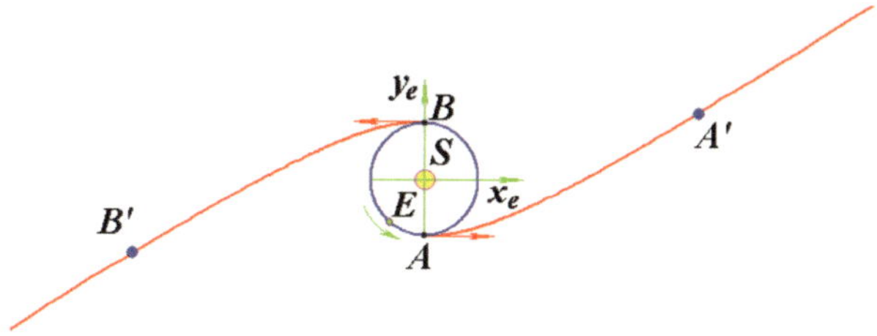

Figure 8.
Trajectories of a triangulation spacecraft for measuring distances to stars: S – *the Sun;* E – *the Earth;* A *and* B – *the launch points of spacecrafts* A' *and* B', *respectively;* $y_e x_e$ – *the plane of the heliocentric ecliptic coordinate system* $x_e y_e z_e$ *for the epoch 2000.0.*

Parameters	Parameter values									
T, years	1.03	2.12	3.08	4.05	5.02	10.1	15.1	20.4	25.4	30.4
r, AU	7	14.4	20	26	32	63	93	125	155	185
D, pc	140	280	400	520	640	1260	1860	2500	3100	3700

Table 5.
The distance D *to stars as determined by triangulation spacecraft, depending on the time of their movement* T *over a distance* r *from the Sun.*

The views of the starry sky seen from the spacecraft in the direction of the z_e-axis and in the opposite direction shall be sent to the Earth at certain time intervals. The view seen from the spacecraft launched at point *A* can be compared with the view of the starry sky seen from spacecraft *B* located at the same distance *r*. This will permit the measurement of distances to objects of $D = 20r/a$ (in parsecs). **Table 5** shows the time of observation *T*, the distance *r* from the Sun in astronomical units, and the distance *D* to astronomical objects, which will be determined using triangulation satellites. After a year of motion, we will be able to reliably know the distance to stars located at a distance $D = 140$ pc; after ten years, at a distance of 1260 pc; and after 30 years, 3700 pc. It should be noted that at a distance of 20 pc, the distance from the Earth will be determined with an error of 20%. Therefore, with an increase of the distance *r* to the spacecraft, it will become possible to refine distances to objects located at distances smaller than the value *D* indicated in **Table 5**.

Range measurements are possible for those distances *r*, up to which the exchange with data between the Space Service on the Moon and the triangulation spacecraft is possible.

Distance *D* to astronomical objects is the basic parameter in astronomy. The sizes of an object, its speed, physical characteristics, and in some cases, it is very physical nature depending on the distance. That is why, in order to be confident in its knowledge of deep space, mankind will always be faced with the task of refining distances to space objects.

Acknowledgements

The data gained in this work were obtained while performing research activities that have been conducted during two decades at the Institute of Earth's Cryosphere, Tyumen SC of SB RAS, Federal Research Center. In recent years this research project has been carried out under Contract Agreement No. 121041600047-2 with RAS. The results are based on the solution of the interaction problem for Solar-system bodies which was obtained using the supercomputers of the Shared-Use Center at the Siberian Supercomputer Center, Institute of Computational Mathematics and Mathematical Geophysics SB RAS, Novosibirsk, Russia. This chapter was read by my son Leonid J. Smulsky and made a number of useful suggestions.

Author details

Joseph J. Smulsky
Institute of Earth's Cryosphere, Tyumen Scientific Centre of SB RAS, Federal Research Center, Tyumen, Russia

*Address all correspondence to: jsmulsky@mail.ru

References

[1] Smulsky JJ. The evolution of the earth's rotational movement for millions of years. The Complex Systems. 2020; **1**(7):3-42

[2] Smulsky JJ. Galactica software for solving gravitational interaction problems. Applied Physics Research. 2012;**4**(2):110-123. DOI: 10.5539/apr.v4n2p110

[3] Smulsky JJ. The system of free access galactica to compute interactions of N-bodies. I. J. Modern Education and Computer Science. 2012;**11**:1-20. DOI: 10.5815/ijmecs.2012.11.01

[4] Smulsky JJ. Future Space Problems and Their Solutions. New York: Nova Science Publishers; 2018. p. 269

[5] Smul'skii II, Krotov OI. Change of angular momentum in the dynamics of the solar system. Cosmic Research. 2015; **53**(3):237-245. DOI: 10.1134/S0010952515020094

[6] Smulsky JJ. Angular momentum due to solar system interactions. In: Gordon O, editor. A Comprehensive Guide to Angular Momentum. New York: Nova Science Publishers; 2019. pp. 1-40

[7] Melnikov V.P. and Smulsky J.J. Astronomical theory of ice ages: New approximations. Solutions and challenges. Novosibirsk: Academic Publishing House; 2009: 84

[8] Smulsky JJ. New geometry of orbital evolution. In: New Geometry of Nature. Kazan': Kazan State University; 2003. pp. 192-195

[9] Smuslky JJ. A mathematical model of the solar system. In: Theoretical and Applied Problems of Nonlinear Analysis. Moscow: Russian Academic Science Dorodnitsyn Computing Center; 2007. pp. 119-139

[10] Garfinkle RA The Earth-Moon System. In: Cognita L. New York: Springer; 2020. 36 p. DOI: 10.1007/978-1-4939-1664-1_2

[11] Smulsky JJ. Dark matter and gravitational waves. Natural Science. 2021;**13**(3):76-87. DOI: 10.4236/ns.2021.133007

[12] Smulsky JJ. Cosmic impacts on the earth and their influence on the Arctic. The Complex Systems. 2017;**4**(25):27-42

[13] Mörner N-A, Editor. Planetary Influence on the Sun and the Earth, and a Modern Book-Burning. Vol. 196. New York: Nova Sciences; 2015

[14] Smulsky JJ, Smulsky Ya J. Dynamic problems of the planets and asteroids, and their discussion. International Journal of Astronomy and Astrophysics. 2012;**2**(3):129-155. DOI: 10.4236/ijaa.2012.23018

Chapter 3

Regolith and Radiation: The Cosmic Battle

Yulia Akisheva, Yves Gourinat, Nicolas Foray and Aidan Cowley

Abstract

This chapter discusses regolith utilization in habitat construction mainly from the point of view of radiation protection of humans on missions of long duration. It also considers other key properties such as structural robustness, thermal insulation, and micrometeoroid protection that all have to be considered in parallel when proposing regolith-based solutions. The biological hazards of radiation exposure on the Moon are presented and put in the context of lunar exploration-type missions and current astronaut career dose limits. These factors guide the research in radiation protection done with lunar regolith simulants, which are used in research and development activities on Earth due to the reduced accessibility of returned lunar samples. The ways in which regolith can be used in construction influence its protective properties. Areal density, which plays a key role in the radiation shielding capacity of a given material, can be optimized through different regolith processing techniques. At the same time, density will also affect other important properties of the construction, e.g. thermal insulation. A comprehensive picture of regolith utilization in habitat walls is drawn for the reader to understand the main aspects that are considered in habitat design and construction while maintaining the main focus on radiation protection.

Keywords: habitat construction, lunar regolith, regolith simulants, ionizing radiation, space radiobiology, radiation protection

1. Introduction

Living on the lunar surface will undoubtedly be a psycho-physical, technological and economical challenge. The main source of protection and support for astronauts will be their habitat. Its construction and design has to offer a counter measure against every stressor exposed onto the crew. While a habitat may be perceived as something static and frozen in the cold of lunar vacuum, it will in fact, in itself become the place of an active battlefield—the battle between radiation and matter, where the health and well-being of the people inside is at stake.

This chapter discusses the utilization of regolith in habitats. Regolith is a local source available in abundance on the lunar surface, which can be relatively easily accessed and collected. Its utilization enables a more sustainable exploration and future settlement. It also reduces the cost of a mission dramatically. However, regolith is a complex material with unique properties that result from space weathering (temperature extremes under vacuum, radiation exposure, micrometeoroid impacts), and the techniques of its utilization and associated technologies are under development and improvement across the global space community. To complicate things further, there is a limited amount of returned lunar samples. In order to satisfy the needs in experimentation, testing and prototyping with regolith, diverse simulants are used. Simulants are specifically designed to resemble the lunar soil in its chemical, mechanical, and thermal properties. Depending on the application, some simulants are perfect replacements of regolith for research and development activities.

The main case under consideration here is regolith for radiation protection of humans. When radiation interacts with matter, it deposits a part of its energy in the target material, produces fragments of nuclei and other secondary emissions. It is important to know how effective regolith is in terms of radiation absorption or attenuation on the one hand, and what kind of secondary particles it will produce on the other. The fact that the radiation environment on the Moon is a diverse mix of particles with different energies and charges makes it complicated to optimize the utilization of regolith for dose reduction. The notion of doses is used to estimate exposure and associated risks. It is always advised to keep the risks and doses to the absolute minimum that is technologically achievable and ethically acceptable. When seeking to reduce doses in space radiation protection, we consider both the doses from primary particles and secondary emissions. In both cases regolith will act as a passive shield, and its constituent molecules will interact with radiations in their unique ways which depend on the mutual chemistry of the projectile-target pair, charge and energy of the incident particle.

As regolith will be the main construction material, it will largely define the thermo-mechanical properties of the habitat wall. It is important to look at the different protective properties in parallel and not dissociate their studies too much. For example, density is crucial for both radiation protection and thermal insulation. A holistic approach to habitat building is discussed here, while keeping the main focus on radioprotection.

The rest of the chapter will introduce lunar habitats, regolith and radiation as the main actors of the cosmic battle. Then, it will outline the problem statement underlining the particular challenges associated with habitat construction on the Moon, regolith utilization, and radiation protection. To fight the problems, the existing armor will be presented. In-situ resource utilization (ISRU) technologies, regolith simulants, and radioprotection techniques will be outlined and discussed. Any good soldier is always on the lookout for more troubles and better solutions. In the context of the cosmic battle it means to be on the lookout for improving ISRU technologies, bettering regolith simulants, and investigating the use and properties of new materials that can either be brought from Earth or made in-situ. A generic conclusion summarizes the main points regarding regolith utilization in habitat construction, mainly from the point of view of radiation protection of astronauts.

1.1 Habitats for long-term exploration

Continuous human presence and surface exploration of the Moon sets an overarching requirement on the lunar habitat that it must sustain human life for several long-term missions and withstand a harsh environment. In other words, the habitat becomes a fortress under a continuous and variable siege of the cosmic and solar radiation, extreme temperatures, and micrometeoroid bombings.

On top of robustness, the habitat must present a comfortable alternative to living on Earth. Working on the Moon for extended periods of time will be extremely challenging, stressful and may even become alienating and daunting. The least that can be done to counteract the psychological burden and physical exhaustion is that the well-being and comfort of astronauts becomes another top-level requirement in habitat construction.

Since the very first steps on the Moon, humanity has been envisioning a long-term presence or even a permanent settlement there. In the most recent years, the global space community focuses primarily on the cislunar space the access to which will enable frequent missions to the surface, ultimately making preparations for the Moon Village [1]. The global exploration roadmap suggests that such efforts should be made in a sustainable way [2]. This leads to the choice of using local materials in habitat construction, and in fact, maximizing their utilization both in hardware and life support.

The most straightforward way to use regolith is to cover a primary structure with it. The primary structure may be brought from Earth, e.g. inflatable or origami-inspired unfolding structure, a metallic cylinder, or even a repurposed part of a spacecraft. **Figure 1** illustrates an artistic view of what such regolith-covered habitats and storage facilities could look like. The authors interpret the image as a capture of the evolution in maturity of ISRU-technology on the Moon. It could be argued that the very first habitats will resemble the one encircled and marked by letter A (in red) since regolith seems to be either loosely piled on top of the structure or compressed and reinforced with dense tiles, which could be produced either through sintering or 3D-printing. Such an approach is feasible at the early stages of exploration. Increasing in complexity, the habitat/storage unit of type B seems to be entirely produced by additive manufacturing. The dark color could be an indicator of another material present in the mixture, e.g. a binder. The surface seems to be rough, possibly owning to the chosen 3D-printing technique which had not yet been thoroughly explored in lunar conditions. Habitat C seems to use more regolith in the material mixture, and it is also produced by additive manufacturing. The triangular and conical shapes observed on the outer layer (both in types B and C) can present a significant advantage in thermal properties of the wall due to partial shadowing—this could help withstand the harsh temperature of the lunar day, which reaches up to 120°C at the equator where the solar heat flux reaches 1300 W/m^2. Finally, the image depicts how the multilayer technology can be utilized with regolith, as the underlying shelter is being covered with another layer of regolith-rich material, seemingly by 3D-printing.

Another straightforward way to benefit from regolith protection is to seek shelter underground. Lava tubes have long been studied as an alternative to living on the Moon, e.g. [3–5]. They extend meters underground and offer a natural

Figure 1.
Solar sintered moonbase, credit: RegoLight Consortium, visualization: LIQUIFER Systems Group.

protection from radiation and micrometeoroids. Most commonly, it is considered that a habitable structure would either be inflated or mounted inside a lava tube. Although it may seem rather convenient and even poetic for the first settlements on the Moon to use the equivalent of caves on Earth, and despite the fact that lava tubes can provide substantial radiation protection (see Section 3.3), this solution has some important limitations which will be outlined in Section 2.1. A surface habitat is considered as the main option for living on the Moon in this chapter.

1.2 Regolith

Committing to a sustainable long-term exploration implies one key material choice—regolith. Abundant on the surface, it will serve as the main force to fight back the cosmic oppressors on the Moon. Regolith, or the lunar soil, will make up the bulk of habitat walls and thus, will act as a shield against incoming radiation particles, heat, and meteoroid projectiles.

Regolith collectively refers to the megaregolith crust consisting of boulders, large particles, grains and powder, or dust. Lunar observations and sample return from the Apollo and Luna missions have resulted in an extensive knowledge of the bulk regolith properties and deciphering some of the history of lunar geology.

Regolith is a complex material. It consists of a mixture of crystalline rock fragments, minerals, breccias, agglutinates, and glasses [6]. Chemical composition of the lunar soil has been thoroughly studied. For radiation protection, it is the most important property as the mutual chemistry of the radiation-matter pair will define the nature of their interactions, and the results in secondary emissions and doses. Two types of regolith are distinguished: mare and highlands, and both are mixes of metallic oxides, dominated by silicon dioxide up to 42–45% in weight [7]. The composition then varies slightly, namely highlands regolith contains more aluminum oxide than the mare type (approximately 25% and 13% respectively [7]). Mare regions contain high levels of titanium dioxide—between 2% and 10% versus the average of 0.5% in highlands soils [7]. It is approximated that the top 30 cm consist of the lunar dust—particles smaller than 100 μm in size with the bulk density of 1.5 g/cm^3 [8]. These loose grains are accessible for collection and utilization in habitat construction. On the Moon, this will make up the majority of ISRU activities.

Currently, the global space sector is investing into its capacity-building related to ISRU technologies [9]. Regolith utilization ranges in ideas from piling-up to sintering, binding with adhesives and 3D-printing. In order to investigate the properties and behavior of raw materials as well as processed products (e.g. regolith bricks), numerical simulations and experiments are carried out. Simulations mainly concern the thermo-mechanical behavior of bulky solids, e.g. how regolith flows and what thermal insulation properties it has. Experiments are usually set up to verify predictions and observe behavior. Humanity currently possesses 382 kg (Apollo program) [10] and 321 g (Luna missions) [8] of lunar regolith from sample return missions, which manifest the greatness of the pioneering efforts in space exploration beyond the Low Earth Orbit (LEO). However, these resources cannot nearly satisfy the global scientific interest and technological demonstration needs in preparation of a lunar outpost. The solution is to simulate the material using its earthly counterparts.

Regolith simulants are like siblings—arguably originating from similar material but having different characteristics. This is due to the fact that simulants are often made to serve different scientific and technical purposes. Literature classifies simulants according to their most prominent properties [11–13]. As such, some are best at simulating mechanical behavior of the lunar soil, and others are almost the exact copies in chemical composition as the returned samples. Continuing the sibling analogy, the differences among regolith simulants may be compared to the different talents that siblings have, which often result from parental investment and resource allocation to activities that nourish those talents.

The first step in working with regolith consists in choosing the appropriate regolith simulant. The main objectives of a habitat are to sustain human life and well-being. Protection from radiation becomes the key player in early habitat planning and regolith simulant considerations as it is one of the main oppressors in the lunar environment. Like under any attack, the forces of resistance must be pulled together. In radioprotective terms, passive shielding is a technique of protection when a material stands in-between a radiation source and the target. The forces of resistance are then the material's nature, or its chemical composition, and areal density. The choice of a passive shield will be based on the most probable radiation-matter interactions, and material optimization will seek to reduce the negative effects of radiation exposure on human health. The interactions between radiation particles and materials produce a diverse variety of results, ranging from energy deposition to nuclear fragmentation and DNA break-down, to mention some. The uniqueness of each interaction originates from the incoming particle's energy, charge and mass. Therefore, the specific radiation environment on the Moon presents a particular challenge to be considered in habitat construction.

1.3 Radiation environment

There are two distinct families of radiation particles: primary and secondary. Primary particles originate from the Sun, our galaxy and distant galaxies [14]. They are high-energy charged particles, mostly protons that move at speeds close to that of light. The diverse mix of ionizing radiation in space, ranging from X-rays to heavy ions with energies up to TeV makes it an extremely challenging environment for radiation protection of humans [14]. When primaries interact with matter, such as the lunar surface, a habitat, or Earth's atmosphere, secondary emissions are produced. The nature and properties of secondary particles depend on the type of interaction that occurred. On the Moon, we can distinguish two branches of secondary emissions: the ones that will occur in the habitat and the lunar neutron albedo.

1.3.1 Galactic cosmic rays (GCR)

Collectively, the particles that make up cosmic radiation are called Galactic Cosmic Rays (GCR). They are baryons (mainly hydrogen protons (83%) and alpha particles, as well as helium (14%) and heavy (1%) nuclei [14]) and electrons that travel in space and are present everywhere. A substantial part of GCR seems to originate from supernova remnants [15, 16] and GCR are believed to be accelerated from outside the Solar System by neutron stars, black holes and supernovae shocks [17]. The mechanism guiding particle acceleration was first proposed by Fermi who explained the energy transfer from magnetized clouds to individual particles [18]. The Fermi I mechanism, also called the diffusive shock acceleration, applied to a strong shock such as from a supernova explosion predicts a power law particle spectrum which has been observed [18].

The magnitude of the GCR spectrum as observed on Earth, and in the rest of the Solar System is correlated with the solar cycle. When the Sun is most active, the enhanced solar magnetic field causes GCR particles to lose some of their energy, and the lower energy particles are affected the most. As such, the fluence of particles of a few GeV/u drops by up to 20% [14]. GCR models account for this relationship with help of the solar modulation parameter [19, 20]. Such models reconstruct the flux of particles mainly from observations, and the most widely used model is the Badhwar-O'Neill (BON) [21]—BON2014 model. Recently, an improved version has been released, BON2020 which reduces model errors largely owning to revised methods of using the solar modulation potential and calibrating free parameters in the local interstellar spectrum for all GCR ions [22].

1.3.2 Solar particle events (SPEs)

The Sun continuously emits particles which make up the solar wind. These are mostly low-energy protons and electrons which are stopped by thin shielding and are thus normally not considered a threat to human space exploration [14]. However during the periods of high activity, the Sun's ejected protons can be accelerated by the chock of a coronal mass ejection or during a solar flare to very high energies. When the energies and flux of the accelerated particles reach high values and extend over a certain period of time, they are registered as Solar Particle Events (SPEs).

SPEs contain mostly protons; include helium ions as well as some highly charged and energetic (HZE) ions. The flux of protons above 30 MeV can exceed 10^{10} cm^{-2} in several hours or days and particles above 50 MeV can penetrate spacesuits and spacecraft [14].

Although SPEs are related to the solar activity and cycle, their appearance remains rather unpredictable [23–25], especially far into the future as exploration-type missions are typically planned. SPEs differ in the prevalent proton energies and particle flux. Some SPEs have been observed and recorded, and data from those events are typically used for radiation protection in space. In 2018, NASA published a report [23] recommending to use the October 1989 series of events as a reference design case for missions beyond LEO, based on SPE storm shelter requirements provided in [26].

1.3.3 Secondary emissions

When primary radiation enters a habitat wall, it reacts with the target molecules and produces secondary emissions. Depending on the nature and energy of the primary-target pair of agents, the produced results will differ from knocked-off electrons to nuclear spallation and formation of ions, neutrons, pions, muons, etc. The most commonly present secondary particles in metallic space vehicles and habitats will be protons of slightly reduced yet still very high energies (when compared to primary protons), neutrons, helium nuclei [27], X and γ rays, and metallic ions of low energies [28]. Some of the secondary emissions (e.g. neutrons) can travel longer in the human body than the primary incoming particle, thus potentially being more harmful. Therefore, secondary emissions must be specifically considered in habitat construction and counter-acted upon, namely in the choice of supplementary materials.

1.3.4 Lunar neutron albedo

Interactions between the primary particles and the lunar soil cause the formation of lunar radiation albedo. It consists mainly of neutrons that are formed from the constant GCR bombardment of regolith and which shoot upwards from the surface. It has been estimated that the neutron albedo can contribute up to 20% of the effective dose on the Moon [17]. Therefore, any human activity on the surface has to take the lunar neutron albedo into account.

2. Problem statement: the battlefield

2.1 Engineering problems: the main aspects to consider

Four main groups of engineering problems have to be considered in habitat construction on the Moon: robustness, feasibility, sustainability and human factors.

Robustness is concerned with the habitat's resistance to structural, thermal and vibro-acoustic loads, meteoroid shocks, and radiation protection. As any house, a habitat has to bear all the loads, some of them present continuously such as the static structural loads, and others appearing occasionally as for example the vibrations from a nearby launch. Meteoroid population around the Moon follows a power law size distribution with small impactors dominating the representation. Traveling at speeds of 3–70 km/s [29], most micrometeoroids are 30–150 μm in size [30]. It has been found that micrometeoroids generally leave impacts of the same order of magnitude as their own sizes [31]. The accumulation of impact craters over time will result in a local density change of the outer shell of the habitat which may affect the mechanical resistance, thermal insulation and radiation protection effectiveness of the structure. Areal density is the most important feature in radioprotective effectiveness of a chosen material. Since all of the main structural stressors will affect the different protective properties of the structure to a greater or lesser extent, they should be considered in parallel when sizing the habitat.

Feasibility considers the technological readiness of the techniques implied in construction as well as cost and power effectiveness of the proposed methods. The mean Technological Readiness Level (TRL) of ISRU technologies reported in the 2021 *In-Situ Resource Utilization Gap Assessment Report* [9] is 3 and the highest TRL is 6 (out of 9). However, these are reported for various uses of regolith to support the human and robotic exploration of the Moon. The TRL of regolith utilization for habitat construction is hard to estimate as only small-scale prototypes of building blocks and techniques have been demonstrated with technologies plausible for lunar utilization [9, 32]. To choose among available ISRU technologies, cost effectiveness and power budget will have to be considered.

Sustainability guides the choice of materials, technologies and techniques in order to ensure a power budget-effective and scalable development and operations of the systems. Maximizing the utilization of local resources is key in achieving sustainable development on the Moon. Regolith will be the main material not only to build but also to operate and maintain facilities. For radiation protection, the degradation of the protective shell over time has to be considered and supported with timely counter-measures. The most important aspect to consider is maintaining the areal density in habitat walls over the years, possibly decades, of exploration.

Human factors regroup such aspects as the crew's mobility, well-being and safety. Surface exploration and accessibility as well as emergency shelters and escape routes have to be considered. Mundane questions such as storage become strategic engineering decisions as storing certain products can locally enhance radiation protection. The choice of the main carrying materials will be mainly guided by their mechanical properties; however the esthetic appreciation is an important factor in habitat design and should not be neglected as supplementary materials will also affect the radioprotective properties of the habitat. An important element among human factors is the visual reference system. Windows are essential in ordinary life, and observations demonstrate how the presence of windows improves human well-being [33, 34]. The fact that astronauts spend a lot of their free time in the Cupola of the International Space Station (ISS) is a clear manifest to that [35].

From a structural point of view, windows are essentially holes that, strictly engineeringly speaking, the structure would be better off without. A window stimulates local concentration of stresses which typically lead to the need of reinforcement. Radiation on the Moon adds another layer to the question of windows: what materials should be used, and how they will affect the radioprotective effectiveness of the habitat.

When the case of lunar lava tubes is considered against the main engineering problems, they evidently score high on feasibility since little preparation is required to use them. However, feasibility is complicated by the need to provide all life support and infrastructure under the ground, possibly extending many meters for ensuring safety. The main consideration regarding robustness is the potential danger of a tube falling in on itself—either upon a meteoroid impact or vibrational excitation (e.g. from a nearby landing/launch). The main show-stopper for lava tubes utilization is surface access and human factors. Humanity seeks to explore the Moon; therefore long surface expeditions are desired. With lava tubes as habitats, astronauts will have to spend a significant amount of time and energy climbing out of their homes onto the surface. For longer expeditions, a surface solar storm shelter must be envisioned to provide immediate protection. In this case, double infrastructure is required, both underground and on the surface, which will largely increase mission's costs and complexity. Most importantly, the psycho-physical effects of living underground on the Moon with no visual reference system, access to natural light or a view of the Earth must be considered. A French "Deep Time" 2021 study [36] has investigated the effects of living in similar conditions on Earth for 40 days; however the lunar case is distinct and more complex due to high levels of stress and alienation which are a part of astronaut life in space.

Most of the engineering problems can be partially answered with regolith utilization. Nevertheless, some additional materials seem inevitable and even desirable—to compensate for certain peculiar behaviors of regolith, thus optimizing material choices for habitat construction.

2.2 Regolith problems: the peculiar behavior of a special material

The lunar soil has been unprotected and constantly bombarded by meteoroids and radiation for several billion years. Such space weathering effects led the material to be crushed and mixed. Particles range in size from a few μm up to a couple of 100 μm, and differ largely in shapes. A distinct property of regolith grains is their extreme adherence and sharpness. These characteristics make regolith uniquely difficult to operate in an effective and safe way. Grains interlock among each other and stick to materials that they come in contact with. They are extremely light, as the average density of a grain is about 3.0 g/cm^3 and most particles measure only a few μm.

A particularly peculiar behavior of regolith on the lunar surface is levitation. Previously considered as the result of meteoritic impacts, particle levitation has recently been tied to the charge buildup from exposure to protons [8]. The difference in charge from the side exposed directly to the solar wind and the side away from the Sun causes charged regolith particles to levitate in attempt to cross the line of difference. This line is the place of the switch between the lunar day and night.

2.3 Radiation problems: the duel of ionizing radiation and radiobiology

In the context of lunar settlement or long-duration missions, astronauts will experience continuous low dose exposure. This type of exposure is higher than that on Earth, which is protected by its magnetosphere and atmosphere, yet it is significantly lower than the single doses delivered as part of radiotherapy. However, some of the radiobiological effects and mechanisms are the same in both cases. Historically, the space sector has been borrowing the findings from radio therapeutic treatments and radiobiology to calculate mission health risks. But space radiation poses important scientific questions about the effects of low doses on cellular and organ levels which can be useful in radio diagnostics and the case of repeated doses.

The so-called absorbed dose, often simply called *dose*, is a measure of energy deposition of a particle in a target material, which is the human tissue in this case. There are several methods to go from energy deposition to the notion of dose which takes into account the biological effects and the harm to organs and the body. Calculating such doses helps to quantify the harmfulness of exposure and cross-compare protective solutions. Based on doses and the associated health risks, which largely depend on the medical history of a person and can be outlined as acute (e.g. nausea) and cumulative short- (e.g. cataracts) and long-term ones (e.g. nervous system function degradation, carcinogenesis), the total mission risks are estimated for astronauts. NASA proposed a model of risk of exposure-induced death (REID) which calculates the risk of death from cancer depending on the age, sex and previous exposure of the astronaut. REID has a hard limit of 3% which means that the total career lifetime exposure should not lead to an increase in the probability of mortality from cancer higher than 3%.

To determine whether a mission is acceptable in terms of radiation exposure, national space agencies set certain exposure limits. As such, there is a short-term limit on 30 day exposure and a career limit, which varies slightly from one agency to another and is also defined by gender in some cases. The former is set by NASA to 250 mSv [37] and the latter averages at 1 Sv across agencies [38]. There is also a specific limit on the exposure to blood-forming organs (BFO). The limit for short-term non-cancer effects is 250 mGy-Eq [37]. Radiation protection solutions must respect these limits and even go above and beyond in looking for dose reduction methods. That is the existing working principle in the context of lunar exploration and settlement, and the global space community is currently putting efforts together to establish specific exploration-type mission limits for joint space activities [38].

3. Existing solutions: the armor

3.1 ISRU technologies

ISRU technologies on the Moon will cover a vast number of activities ranging from collection, storage, manipulation, recycling, treatment, and post-processing. Regarding habitat construction, it should be noted that first, the construction area needs to be cleared, leveled and compacted to control the spread of lunar dust. Then, such an area can be used for building a habitat.

Currently, the global space community investigates sintering, molding, brick-making, and 3D-printing with regolith. The techniques require different types of expertise, machinery, level of automation/human presence, power, and supplementary materials. The readiness levels of the technologies varies drastically as some techniques have been investigated for a number of decades while other started to gain a significant level of industrial and engineering interest in more recent years. As such, the idea of piling up loose regolith dates back to the Apollo era, cement and concrete production has been investigated since the 1980s [32], and additive manufacturing has been attracting a lot of attention in the last tens of years.

3.2 Regolith simulants

Typically simulants are made by crushing down terrestrial rocks of basaltic origins that largely resemble the chemical composition of the rock component of the lunar soil. The mixture can be improved by adding any particular minerals or glasses, as was done for the very first lunar regolith simulant JSC [13] when knowledge about lunar soils advanced thanks to sample return.

Including both mare and highlands types, there are a few tens of simulants that are being produced and used across academia and industry. These simulants respond to different engineering and scientific needs, and are used in technological demonstrations and experiments. In a user guide [11], NASA suggests that particle composition, size distribution, shape distribution and bulk density are the most important properties in a regolith simulant. Indeed, these factors will largely define the thermo-mechanical and chemical properties of the raw material and also outline how it will interact with its environment (e.g. static charge) and other materials (e.g. abrasive nature of the material). For radiation protection purposes, chemical composition and areal density are key factors that will define the effectiveness of regolith shielding. Radiation cross-sections are calculated from molecular formulas and are used to predict the interactions between the incoming radiation and matter. Areal density in g/cm^2, measures how much passive shielding is present in the way of the incident particles. Simply put, in dense materials where molecules sit closely together, there is a higher chance for an incoming primary particle to interact either with the nucleus or the electrons of the molecules. Bulk density in g/cm^3, defines whether and how much the simulant needs to be compressed in order to reach the areal density required for radiation protection.

3.3 Radiation protection

Deviation, distance, time, counter measures, and materials are the only units to put forward at the front line against radiation. In a lunar habitat however, large-scale particle deviation is not a feasible option. Increasing the distance to radiation source in space is impossible as the primary particles are omnipresent in interstellar space, and reducing time exposure may be in conflict with the scientific and exploratory missions' objectives. Although biological counter measures are currently being explored, this research is in its early stages and it is further challenged by individual responses to repeated exposures and hyper sensitivity to low doses. This leaves it to the strategic choice of passing shielding to protect astronauts from radiation. The best choice consists in the material that will absorb the maximum amount of primary radiation while producing the least amount of secondary emissions. The complexity and diversity of the space radiation environment makes this choice all the more difficult. However due to the large shipment costs to the Moon and the abundance of loose regolith on the surface, it becomes the main shielding material in a habitat.

As most units do, the radiation protection community has a guiding motto—a principle proposed by NASA—As Low As Reasonably Achievable (ALARA). It pushes the community to engineer ways to bring down the organ and whole-body doses, ultimately aiming at lower health risks associated with exposure.

As outlined in previous sections, one possible way to maximize radiation protection on the Moon is to build a habitat underground. Studies [3, 39] suggest that several meters under the surface, GCR exposure levels become comparable to those on Earth—a few mSv/year. However due to the major drawbacks of using and living in lava tubes expressed in Section 2.1, this option is not considered for an early settlement here. However, lava tubes should be investigated further for the potential use as shelters from SPEs.

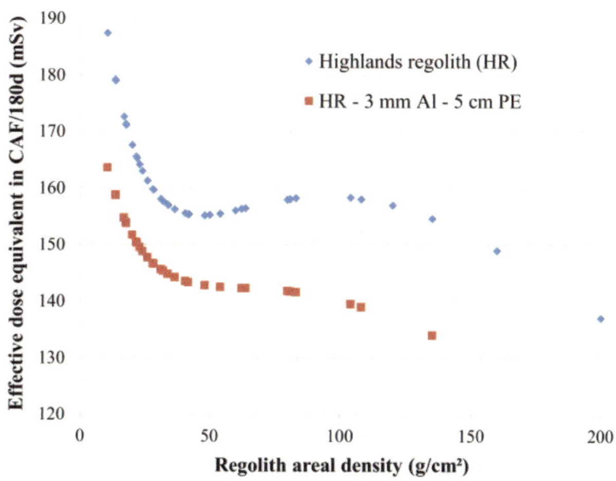

Figure 2.
Effective dose equivalent from GCR in CAF/180 days behind highlands regolith (HR) and multilayer shielding (HR—3 mm aluminum, Al—5 cm polyethylene, PE) as a function of regolith areal density. Based on results in [39].

The best way to optimize passive shielding is to utilize the most effective molecules in terms of radiation protection. Extensive studies [23, 40, 41] show that low atomic mass materials act best as shielding against heavy ions and high-energy protons as they present more nuclei in the path of the incoming particles, thus maximizing the stopping power for the same shield thickness in mass per unit area, if compared to heavier atomic mass counterparts. The top sergeant in this respect is protium or hydrogen ^{1}H because on top of its low atomic mass, it contains no neutrons and thus its utilization enables to bring down the secondary neutron production.

When the choice of chemistry of the main shielding is done or limited, the two cards left to play are areal density (of regolith in the lunar case) and the combination of supplementary materials which can be brought from Earth in moderate amounts, or possibly fabricated in-situ in the future. The term *combination* here includes both the types of added materials (their chemistry, density, H-richness, etc.) and the order or composition of the different layers together. Dense materials present a higher probability for primary radiation to interact with the target molecules since they sit tightly together. Therefore compression and sintering techniques are investigated with regolith and regolith simulants. If compressed regolith is complemented by low-atomic mass or hydrogen-rich materials in a multilayer structure then ALARA principle may be approached. This effect is illustrated in **Figure 2** which summaries the results of a deterministic study [39] with highlands regolith and a multilayer of regolith, aluminum and polyethylene, the latter being rich in hydrogen. The results demonstrate how the addition of aluminum and especially polyethylene leads to a significant reduction in the effective (whole-body) dose equivalent in a Computerized Anatomical Female (CAF) model in lunar GCR environment, simulated in OLTARIS [42] using the BON2014 model.

4. On the lookout for new solutions

To follow the ALARA principle implies to be on the lookout for material enhancements, new materials, and the evolution of ISRU technology. Starting with an evaluation of commonly used materials, it is wise to look into possible combinations of those with regolith. As such, a study of 59 space materials [40] concluded that polymers should be used instead of metals in space where possible. In parallel, polymer 3D printing and sintering techniques with regolith are being developed (e.g. [43, 44]).

Besides the development of new materials, the utilization of multilayered structures is being investigated. The use of multiple layers of different complementary materials is not a new concept is space, as it has been used since the very first days of exploration, in particular in Extravehicular Mobility Units (EMUs). However, the radioprotective properties of such commonplace materials as Kevlar in EMUs has only been investigated recently [45]. The ROSSINI study [45] performed accelerator-based tests of several multilayers with He, F, and C beams of 1000, 962–972, and 430 MeV/u respectively. Among other, it concluded that the addition of LiH to a Moon regolith simulant enhanced protection from radiation by up to 20%. However, any such study is limited to the particular energy and type of primary particles. Overall recommendations require further tests and consideration of secondary emissions—especially neutrons [45].

5. Conclusion

The unpredictability of solar behavior is being anticipated and compensated for with large margins for error, where no error is accepted. Models of GCR are being improved to provide more precise calculations of doses and associated risk estimations. New technologies, experiments, measurements, materials, and simulation models are being developed and tested. Observational missions, such as those to Lagrange points (e.g. ESA missions [46] and NASA's DSCOVR mission [47]) are aimed at providing fast capabilities of forecasting and alerting. All these elements make an intellectual playground for radiation protection engineers and scientists—to make a safe ground for living on the Moon.

Guided by the best practices and prioritizing human comfort and well-being, the specialists on Earth will be making crucial choices for those who will go to the Moon. Under the assumption that habitats are to remain highly effective and functional for several astronaut generations to come, global and diversified efforts are required to design, qualify and supervise their construction. Habitat construction working groups are expected to incorporate wide research expertise, originating from fields such as radiobiology, medicine, aerospace engineering, mining, construction, architecture, material sciences, etc. In cooperation, such groups are better equipped to challenge the stressors of the lunar environment.

Acknowledgements

The authors would like to express gratitude for offered financial support and expertise that enabled the publishing of this chapter. Gratitude is extended to the European Space Agency, TRAD Tests & Radiations, Inserm the National Institute of Health and Medical Research, and the Department of Mechanics, Structures and Materials of ISAE-SUPAERO, l'Institut Supérieur de l'Aéronautique et de l'Espace. Specifically, the authors would like to thank Dr. Advenit Makaya from ESA and the experts from TRAD Tests & Radiations for their expertise and implication in the project called Protective Use of Regolith for Planetary and Lunar Exploration (PURPLE), which is the framework under which this chapter was developed. At ISAE-SUPAERO, the authors would like to thank V. Godivier and X. Foulquier.

Nomenclature

C	carbon
eV	electron volt
F	flourine
G	giga
Gy-Eq	gray-equivalent
H	hydrogen
He	helium
LiH	lithium hydride
m	milli
M	mega
Sv	Sievert
T	terra
u	atomic mass unit
X	X ray
γ	gamma ray

Author details

Yulia Akisheva[1*], Yves Gourinat[2], Nicolas Foray[3] and Aidan Cowley[4]

1 Institut Supérieur de l'Aéronautique et de l'Espace (ISAE-SUPAERO), Toulouse, France

2 ISAE-SUPAERO, Toulouse, France

3 Inserm U1296 Unit "Radiation: Defence, Health, Environment"—Centre Léon-Bérard, Lyon, France

4 European Space Agency (ESA), European Astronaut Centre (EAC), Köln, Germany

*Address all correspondence to: yulia.akisheva@isae-supaero.fr

References

[1] Köpping Athanasopoulos H. The moon village and space 4.0: The 'Open Concept' as a new way of doing space? Space Policy. 2019;**49**:101323. DOI: 10.1016/j.spacepol.2019.05.001

[2] Laurini KC, Hufenbach B, Hill J, Ouellet A. The global exploration roadmap and expanding human/robotic exploration mission collaboration opportunities. Jerusalem, Israel: Presented at the 66th International Astronautical Congress, International Astronautical Federation; 2015. p. 9

[3] Angelis GD, Wilson JW, Clowdsley MS, Nealy JE, Humes DH, Clem JM. Lunar lava tube radiation safety analysis. Journal of Radiation Research. 2002;**43**(S):S41-S45. DOI: 10.1269/jrr.43.S41

[4] Benaroya H, Bernold L. Engineering of lunar bases. Acta Astronautica. 2008;**62**(4-5):277-299. DOI: 10.1016/j.actaastro.2007.05.001

[5] Haruyama J et al. Lunar holes and lava tubes as resources for lunar science and exploration. In: Badescu V, editor. Moon. Berlin, Heidelberg: Springer Berlin Heidelberg; 2012. pp. 139-163. DOI: 10.1007/978-3-642-27969-0_6

[6] Heiken GH, Vaniman DT, French BM. Lunar Sourcebook: A User's Guide to the Moon. Cambridge: Cambridge University Press; 1991

[7] Stoeser D, Wilson S, Rickman D. Design and Specifications for the Highland Regolith Prototype Simulants NU-LHT-1M and -2M. p. 24

[8] Colwell JE, Batiste S, Horányi M, Robertson S, Sture S. Lunar surface: Dust dynamics and regolith mechanics. Reviews of Geophysics. 2007;**45**(2): RG2006. DOI: 10.1029/2005RG000184

[9] International Space Exploration Coordination Group. In-Situ Resource Utilization Gap Assessment Report. 2021

[10] Lunar Rocks and Soils from Apollo Missions. Available from: https://curator.jsc.nasa.gov/lunar/ [Accessed: October 12, 2021]

[11] Schrader CM, Systems B, Rickman DL, McLemore CA, Fikes JC. Lunar Regolith Simulant User's Guide. NASA/TM—2010-216446. National Aeronautics and Space Administration; 2010

[12] He C. Geotechnical Characterization of Lunar Regolith Simulants. p. 241.

[13] Taylor LA, Pieters CM, Britt D. Evaluations of lunar regolith simulants. Planetary and Space Science. 2016;**126**: 1-7. DOI: 10.1016/j.pss.2016.04.005

[14] ICRP. Assessment of radiation exposure of astronauts in space. ICRP Publication 123. Ann. ICRP; 2013;**42**(4)

[15] Blasi P. Origin of Galactic Cosmic Rays. Nuclear Physics B. Proceedings Supplements. 2013;**239-240**:140-147. DOI: 10.1016/j.nuclphysbps.2013.05.023

[16] Cronin JW. Cosmic rays: The most energetic particles in the universe. Cosmic Rays. 1999;**71**(2):8

[17] Dachev TP et al. An overview of RADOM results for earth and moon radiation environment on Chandrayaan-1 satellite. Advances in Space Research. 2011;**48**(5):779-791. DOI: 10.1016/j.asr.2011.05.009

[18] Amato E. The origin of galactic cosmic rays. International Journal of Modern Physics D. 2014;**23**(07):1430013. DOI: 10.1142/S0218271814300134

[19] Potgieter MS. Solar modulation of cosmic rays. Living Reviews in Solar Physics. 2013;**10**(1):3. DOI: 10.12942/lrsp-2013-3

[20] Usoskin IG, Bazilevskaya GA, Kovaltsov GA. Solar modulation parameter for cosmic rays since 1936 reconstructed from ground-based neutron monitors and ionization chambers: Cosmic ray modulation. Journal of Geophysical Research: Space Physics. 2011;**116**(A2). DOI: 10.1029/2010JA016105

[21] O'Neill PM. Badhwar-O'Neill galactic cosmic ray model update based on advanced composition explorer (ACE) energy spectra from 1997 to present. Advances in Space Research. 2006;**37**(9):1727-1733. DOI: 10.1016/j.asr.2005.02.001

[22] Slaba TC, Whitman K. The Badhwar-O'Neill 2020 GCR Model. Space Weather. 2020;**18**(6):1. DOI: 10.1029/2020SW002456

[23] Norbury JW et al. Advances in space radiation physics and transport at NASA. Life Sciences and Space Research. 2019;**22**:98-124. DOI: 10.1016/j.lssr.2019.07.003

[24] Kim M-HY, De Angelis G, Cucinotta FA. Probabilistic assessment of radiation risk for astronauts in space missions. Acta Astronautica. 2011;**68** (7-8):747-759. DOI: 10.1016/j.actaastro.2010.08.035

[25] Kim M-HY, Hayat MJ, Feiveson AH, Cucinotta FA. Prediction of frequency and exposure level of solar particle events. Health Physics. 2009;**97**(1): 68-81. DOI: 10.1097/01.HP.0000346799.65001.9c

[26] Townsend LW et al. Solar particle event storm shelter requirements for missions beyond low Earth orbit. Life Sciences and Space Research. 2018;**17**:32-39. DOI: 10.1016/j.lssr.2018.02.002

[27] Peracchi S et al. Modelling of the silicon-on-insulator microdosimeter response within the International Space Station for astronauts' radiation protection. Radiation Measurements. 2019;**128**:106182. DOI: 10.1016/j.radmeas.2019.106182

[28] Ferlazzo M, Devic C, Foray N. "Premiers éléments de radiobiologie spatiale." Unpublished manuscript, Inserm U1296 Unit "radiation: defence, health, environment," 2021

[29] Allende MI, Miller JE, Davis BA, Christiansen EL, Lepech MD, Loftus DJ. Prediction of micrometeoroid damage to lunar construction materials using numerical modeling of hypervelocity impact events. International Journal of Impact Engineering. 2020;**138**:103499. DOI: 10.1016/j.ijimpeng.2020.103499

[30] Grun E, Zook HA, Fechtig H, Giese RH. Collisional balance of the meteoritic complex. Icarus. 1985;**62**: 244-272. DOI: 10.1016/0019-1035(85)90121-6

[31] Holsapple KA. The scaling of impact processes in planetary sciences. Annual Review of Earth and Planetary Sciences. 1993;**21**:333-373

[32] Isachenkov M, Chugunov S, Akhatov I, Shishkovsky I. Regolith-based additive manufacturing for sustainable development of lunar infrastructure—An overview. Acta Astronautica. 2021;**180**:650-678. DOI: 10.1016/j.actaastro.2021.01.005

[33] Chang C et al. Life satisfaction linked to the diversity of nature experiences and nature views from the window. Landscape and Urban Planning. 2020;**202**:103874. DOI: 10.1016/j.landurbplan.2020.103874

[34] Elsadek M, Liu B, Xie J. Window view and relaxation: Viewing green space from a high-rise estate improves urban dwellers' wellbeing. Urban Forestry & Urban Greening. 2020;**55**:126846. DOI: 10.1016/j.ufug.2020.126846

[35] White WF. The overview effect and creative performance in extreme human environments. Frontiers in Psychology. 2021;**12**:584573. DOI: 10.3389/fpsyg.2021.584573

[36] Associated Press. French Isolation Study for 15 People Ends After 40 Days in Cave. VOA. Available from: https://www.voanews.com/a/science-health_french-isolation-study-15-people-ends-after-40-days-cave/6205014.html [Accessed: October 28, 2021]

[37] National Aeronautics and Space Administration. NASA space flight human-system standard volume 1, revision A: Crew health. NASA-STD-3001. 2014

[38] Walsh L et al. Research plans in Europe for radiation health hazard assessment in exploratory space missions. Life Sciences and Space Research. 2019;**21**:73-82. DOI: 10.1016/j.lssr.2019.04.002

[39] Akisheva Y, Gourinat Y. Utilisation of Moon regolith for radiation protection and thermal insulation in permanent lunar habitats. Applied Science. 2021;**11**(9):3853. DOI: 10.3390/app11093853

[40] Bond DK, Goddard B, Singleterry RC, León SB y. Evaluating the effectiveness of common aerospace materials at lowering the whole body effective dose equivalent in deep space. Acta Astronautica. 2019;**165**:68-95. DOI: 10.1016/j.actaastro.2019.07.022

[41] DeWitt JM, Benton ER. Shielding effectiveness: A weighted figure of merit for space radiation shielding. Applied Radiation and Isotopes. 2020;**161**:109141. DOI: 10.1016/j.apradiso.2020.109141

[42] Singleterry RC et al. OLTARIS: On-line tool for the assessment of radiation in space. Acta Astronautica. 2011;**68**(7-8):1086-1097. DOI: 10.1016/j.actaastro.2010.09.022

[43] Montes C et al. Evaluation of lunar regolith geopolymer binder as a radioactive shielding material for space exploration applications. Advances in Space Research. 2015;**56**(6):1212-1221. DOI: 10.1016/j.asr.2015.05.044

[44] Sik Lee T, Lee J, Yong Ann K. Manufacture of polymeric concrete on the Moon. Acta Astronautica. 2015;**114**:60-64. DOI: 10.1016/j.actaastro.2015.04.004

[45] Giraudo M et al. Accelerator-based tests of shielding effectiveness of different materials and multilayers using high-energy light and heavy ions. Radiation Research. 2018;**190**(5):526. DOI: 10.1667/RR15111.1

[46] Monitoring space weather. Available from: https://www.esa.int/Safety_Security/Monitoring_space_weather2 [Accessed: October 15, 2021]

[47] DSCOVR: Deep Space Climate Observatory, NESDIS. Available from: https://www.nesdis.noaa.gov/current-satellite-missions/currently-flying/dscovr-deep-space-climate-observatory [Accessed: October 15, 2021]

Section 3

Habitat and Humans

Chapter 4

Modeling Radiation Damage in Materials Relevant for Exploration and Settlement on the Moon

Natalia E. Koval, Bin Gu, Daniel Muñoz-Santiburcio and Fabiana Da Pieve

Abstract

Understanding the effect of radiation on materials is fundamental for space exploration. Energetic charged particles impacting materials create electronic excitations, atomic displacements, and nuclear fragmentation. Monte Carlo particle transport simulations are the most common approach for modeling radiation damage in materials. However, radiation damage is a multiscale problem, both in time and in length, an aspect treated by the Monte Carlo simulations only to a limited extent. In this chapter, after introducing the Monte Carlo particle transport method, we present a multiscale approach to study different stages of radiation damage which allows for the synergy between the electronic and nuclear effects induced in materials. We focus on cumulative displacement effects induced by radiation below the regime of hadronic interactions. We then discuss selected studies of radiation damage in materials of importance and potential use for the exploration and settlement on the Moon, ranging from semiconductors to alloys and from polymers to the natural regolith. Additionally, we overview some of the novel materials with outstanding properties, such as low weight, increased radiation resistance, and self-healing capabilities with a potential to reduce mission costs and improve prospects for extended human exploration of extraterrestrial bodies.

Keywords: space radiation, multiscale modeling, defects, semiconductors, alloys, composites, solar cells, habitat on the Moon

1. Introduction

Preparing for life on another planet or a planetary object requires an enormous effort from scientists and engineers [1]. The first steps toward extraterrestrial life are the crewed missions to the Moon, aiming to build the basis for the future long-term presence of humans beyond Earth. A remarkable amount of research and feasibility studies are being done by the European Space Agency (ESA) in Europe [2] and the National Aeronautics and Space Administration (NASA) in the USA [3, 4] on how to construct a "new home in space," in a manner to eliminate the need for supply materials from Earth.

In this context, the use of space resources is one of the key directions in preparation for future human missions to the Moon. The so-called *in situ* Resource Utilization (ISRU) program by ESA and NASA explores the possibility of converting local resources of space bodies into valuable products and materials [5–8]. ISRU will ensure the sustainability and energy efficiency of space exploration, reduce the cost of delivery from Earth, and minimize mission risks. Among the topics of current ISRU research are producing metals and construction materials by transforming local regolith and rocks [9, 10], harvesting oxygen and hydrogen from minerals and water [10, 11], and growing plants [12, 13]. In this sense, the development of structurally sound composite materials with superior properties that can benefit from ISRU is crucial for preparing missions to the Moon. Many aspects of habitat construction, from large-scale infrastructure (e.g., communication and energy generation and storage) to manufacturing (e.g., equipment, tools, and machinery), would benefit from ISRU.

In space and on the lunar surface, there are many factors potentially leading to damage in materials, such as exposure to vacuum, extreme thermal conditions, impact collisions with micrometeoroids, and radiation [14]. Among these, radiation is considered particularly harmful for different functional components and instruments of spacecraft and lunar surface missions. Radiation can induce structural defects that evolve from nanoscale to micro- and macro-damage, causing degradation of the mechanical, thermal, and electrical properties of materials or can even lead to direct failure in electronic signals before interacting with the very structural composition of the material. Therefore, improving the radiation resistance of materials to be used in space missions and searching for more radiation-resistant materials is of utmost importance. The research effort is directed toward finding composite materials that can better withstand radiation and other challenges faced by mission components in space and on space bodies and exhibit self-healing capabilities [15].

In this chapter, we first introduce some relevant materials for two of the most critical applications on the Moon, i.e., habitat construction and energy production. Then, we provide an overview of the radiation environment on the lunar surface and different radiation effects that can be induced in materials by such an environment. We then discuss the ways of combining traditional methods commonly used to study radiation effects with recent advanced approaches in materials modeling and provide examples of radiation-effects modeling studies on different materials. Additionally, we discuss the possibilities of using novel promising materials with exceptional properties relevant for space exploration, with an emphasis on their radiation resistance.

2. Materials for practical applications on the Moon

NASA has identified the most important components of the lunar mission as (i) design and construction of habitats and (ii) resource and power management [16]. In particular, the emphasis is on lightweight materials that will be critical for mass reduction and thus increase the science return of the mission. Both components mentioned above will strongly rely on ISRU, i.e., *in situ* regolith processing and recycling [8, 17]. Below we provide examples of materials that will be of use for both habitat construction and power generation.

2.1 Materials for habitats

Constructing a habitat on the Moon can be done in two ways, by delivering materials from Earth and by using local resources. Although the latter option is more sustainable, the first one cannot be completely avoided. An important consideration that needs to be made when choosing materials is the type of habitat. NASA considers several types of habitat for different use, namely rigid (metals, alloys, and concrete) [18], inflatable (e.g., inflatable concrete [19]), or hybrid structures, as well as underground construction [20]. Depending on the type of habitat, different materials will be used [16, 21]. For example, unprocessed lunar regolith may be used for radiation shielding of habitat (e.g., lunar regolith geopolymer) [22–25], as well as for construction when converted into concrete [26, 27], 3D-printed [28–30], or processed into other construction material (e.g., bricks and glass) [16, 21]. For materials delivered from Earth, it is crucial to ensure their low weight, as well as resistance to very high and very low temperatures (which change from 127°C in the daytime to −173°C at night on the Moon surface) and radiation, durability, reusability, and structural reliability [16].

Metals and alloys are essential structural materials for construction given their compressive strength and good tensile properties and for other applications, such as energy carrier/storage (wires) [31] or equipment (e.g., excavation tools, molds, and rovers) [32]. Al, Ca, Fe, Ti, and Mg are the most abundant metals in the lunar regolith, which also contains smaller amounts of Ni, Cr, Mn, Zr, and V [5, 20]. These metals—together with Si, also abundant on the Moon—can be used to produce alloys. However, only Fe can be easily separated from regolith (using magnets). Other metals are present in the form of oxides and thus have to be obtained by manufacturing. Metal and alloy manufacturing will be extremely important for the exploration of the Moon as they represent an essential part of the construction and are critical ingredients for most technologies.

2.2 Materials for energy production

One of the crucial steps toward the Moon exploration and settlement is a reliable energy technology for electricity generation and power storage [33, 34] that would withstand the temperature gradients, high levels of radiation, and impact. The primary energy sources considered for future crewed lunar missions are solar power [35, 36], nuclear power [37], and fuel cells [38, 39]. Other ways may include the production of electricity from the excess heat from the sunlight collected by an "evergreen" inflatable dome [40]. In this chapter, we focus on solar cells, a safe and reliable source of electricity in space.

In the past decades, solar cells for space applications have evolved from single-crystalline Si-based cells to multi-junction (MJ) ones based on GaInP, GaAs, and Ge [41–43]. A promising class of materials for next-generation lightweight and high-power-conversion efficiency [44] solar cells are hybrid organic-inorganic perovskites (HOIPs) [45–47], which are considered as potential candidates for use on future lunar bases [34].

HOIPs possess a unique combination of properties, such as enhanced charge carrier mobility [48–51], diffusion length, and lifetime [48, 52, 53], high optical absorption [54, 55], and low production costs [56], representing a paradigm shift in solar cell technology [57] on Earth [58] and for space applications [59–62]. Given their flexibility [63], low weight, small dimensions (0.5 μm as compared to 200 μm for Si solar cells), the possibility of *in situ* manufacturing via 3D-printing techniques

[60, 64, 65] at low temperature, and their high resistance to radiation [60, 66–71], HOIPs qualify as exceptional candidates for easily deployable and resilient solar cells in space missions.

3. Radiation environment on the Moon and its effect on materials

3.1 Radiation environment on the lunar surface

The radiation environment on the Moon is constituted, apart from solar electromagnetic radiation, by three radiation "populations"—the constant solar wind, the intense but sporadic Solar Energetic Particles (SEPs), and the constant background of Galactic Cosmic Rays (GCRs). A summary of the radiation environment on the lunar surface is given in **Table 1**.

The solar wind is a constant flux of plasma from the upper atmosphere of the Sun. It consists mainly of ionized hydrogen (protons and electrons), a small percentage of α-particles, and trace amounts of heavier ions, with kinetic energy between 0.5 and 2 keV/nucleon [75]. The solar wind flux, temperature, density, and speed vary over time and solar longitude and latitude. The lunar surface is under continuous bombardment by the solar wind, as the Moon does not have a significant global geomagnetic field that could deflect solar particles. Particles penetrate the surface and undergo collisions with the ions of the lunar regolith. Their penetration depth depends on the impact energy, angle of incidence, and composition of the target surface. For a proton with a nominal energy of 1 keV, the penetration depth is typically about 20 nm [79]. The implanted protons diffuse and chemically combine with the regolith atoms, such as oxygen, or become trapped in physical defects. Recent studies have suggested that the implantation of solar wind protons in the lunar regolith is a major source of hydrogen in the formation of OH/H_2O [80, 81], whose presence is confirmed by experimental measurements [82].

SEPs originate from solar transient events, such as coronal mass ejections or flares, and consist in a sudden intense flux of high-energy protons and electrons (and a small amount of α-particles and heavier ions) [76, 78]. Typical energies of SEPs range from ten to hundreds of MeV. Such transient events have a higher occurrence probability during solar maximum, but they may also occur during solar minimum. Studies have shown that the lunar surface can charge to a high negative potential up to a few kV during SEP events [83, 84]. Such values of the potentials are much higher than the typical night-side potentials of a few hundred volts negative and may increase the risk of electrostatic discharge. The latter represents an additional hazard to the already dangerous radiation environment on the lunar surface.

GCRs constitute the slowly varying, low-intensity (few particles/cm^2(m^2) per second), highly-energetic radiation background in space. They are mainly

Source	Particles	Energy, MeV/nuc	Flux, $nuc/cm^2/s$
Solar Wind	Protons & electrons ~95%, α-particles ~4%, heavy ions ~1%	$\sim 10^{-3}$	$\sim 10^8$
SEPs	Protons >90%, electrons, α-particles, heavy ions <1%	$\sim 1–10^2$	$0 - 10^6$
GCRs	Protons ~ 87%, α-particles ~12%, heavy ions ~1%	$\sim 10^2–10^4$	2–4

Table 1.
Radiation particle types, their flux, and energies on the lunar surface [72–78].

associated with supernova explosions in the galaxy, but extra-galactic contributions also exist. GCRs are constituted by ~87% of hydrogen ions (protons), 12% of α-particles, 1–2% of high-energy and highly charged ions (high-charge Z and energy (HZE)-particles), and 1% of electrons and positrons [85]. The energy spectrum of GCRs covers a wide range, extending roughly up to 10^{18} eV, with higher energies (up to 10^{21} eV) being associated with ultrahigh-energy GCRs originating from extra-galactic sources. GCRs are modulated by the heliospheric field linked to solar activity. At solar maximum, the solar magnetic field increases, shielding the heliosphere from the lowest energy component of GCRs [86], thus decreasing the overall GCRs flux. At the solar minimum, the reduced solar magnetic field leads to a more intense GCRs flux in our interplanetary space [87, 88].

The annual exposure caused by GCRs on the lunar surface is ~380 mSv during solar minimum and ~110 mSv during solar maximum, as compared to the annual dose of natural ionizing radiation of 2.4 mSv on Earth [89] (1 Sv—1 Sievert, represents the equivalent biological effect of the deposit of a Joule of radiation energy in a kilogram of human tissue). The worst-case scenario studies suggest that SEPs may lead to a much higher exposure of ~1 Sv or even reach > 2 Sv per event [90]. Studies of the radiation dose of GCRs and SEPs at the lunar surface and in a lava tube [90, 91] have shown that the exposure may be reduced to values similar to Earth in horizontal lava tubes.

3.2 Radiation-induced effects in materials

The effects of radiation on materials and devices can be cumulative (long term) and noncumulative (caused even by a single particle). The so-called Single Event Effects (SEEs) can occur when an ionizing particle passing through an electronic device carries a charge large enough to affect the device's performance. SEEs in aerospace technology can lead to errors, corrupt the data, create noise, reset the device, or even cause fatal part failure [92–95]. Cumulative radiation damage, on the other hand, occurs through continuous radiation exposure or exposure to intense flux due to SEPs events and can lead to the degradation of optical components and solar cells, eventually causing permanent damage. The total ionizing dose experienced by an electronic device can cause variations in threshold voltage or leakage current.

Cumulative non-ionizing damage in materials due to protons, electrons, and neutrons (originating from the interaction of energetic protons and electrons with the lunar surface) leads to defect formation (displacement damage) [94]. The types and sources of radiation, as well as the effects it can cause in materials, are summarized in **Table 2**.

Particle type	Energy	Sources	Radiation effects
Electrons	> 1 MeV	SEPs	Ionization radiation damage
Protons	0.1 − 1 MeV	SEPs	Surface damage to materials
Protons	1 − 10 MeV	SEPs accelerated in shocks	Displacement damage in solar cells
Protons	> 10 MeV	SEPs and GCRs	Ionization and displacement damage, background counting in sensors
Protons	> 50 MeV	SEPs and GCRs	Single event effects
Ions	> 10 MeV/nuc	SEPs and GCRs	Single event effects

Table 2.
Sources and types of radiation and the effects it causes in materials and devices [96].

Cumulative radiation damage is a multiscale process in terms of time and length. A schematic representation of the so-called displacement damage cascade is shown in **Figure 1**. At first, an energetic external particle approaches (**Figure 1**(1)) and enters the target (**Figure 1**(2)). As the particle passes through the material, it first transfers its kinetic energy to electronic degrees of freedom of the target (electronic stopping) (**Figure 1**(3)). Electronic excitations happen at a very short time scale (~100 as). After the particle has been slowed down by the target's electrons, it undergoes nuclear elastic collisions, displacing atoms in the target (Primary Knock-on Atoms, PKAs) that constitute themselves additional projectiles (**Figure 1**(4)). The PKA collides with other atoms creating a cascade of collisions [97] (**Figure 1**(5)). Atomic displacements induce the creation of different types of point defects, such as vacancies and interstitials (Frenkel pairs) and defect clusters (**Figure 1**(6)) and happen on a much longer time scale (up to ns). Eventually, many defects are healed due to the thermal motion of atoms (annealing stage, **Figure 1**(7)), leaving a finite number of defects in the structure (**Figure 1**(8)).

Atomic displacements described above lead to defect clustering and eventual amorphization in crystalline materials. Consequently, mechanical, physical, and other properties of the irradiated material can be significantly altered. The scale of the changes depends on the energy of incoming particles and the actual number and spatial distribution of survived defects after eventual self-healing [98].

The radiation-induced effects after atomic displacements strongly depend on the type of material. For metals and metallic alloys, the main effect of radiation is the generation of dislocation loops and point defects which cause significant radiation-induced strengthening or hardening. As a result, the ductility and fracture toughness of the metals (alloys) can be reduced, leading to brittle behavior [99]. Ductile-to-brittle transition is especially pronounced at low temperatures at which the defect mobility, and consequently the annealing of defects, is reduced.

As to other materials, such as semiconductors in solar cells, cumulative exposure to space radiation or high SEPs fluxes can strongly affect the performance of MJ solar cells [100]. Moreover, the impacting radiation can reduce the transmittance of the protective SiO_2 cover-glass on top of MJ cells by inducing color centers in the oxide material. The color centers appear when electrons excited by radiation become trapped by impurities in the oxide to form stable defect complexes. On the other hand, the radiation which is not blocked by the cover-glass causes damage in the functional layers of MJ solar cells by displacing atoms. Different energy levels can be created within the bandgap as a consequence of such structural defects. Such

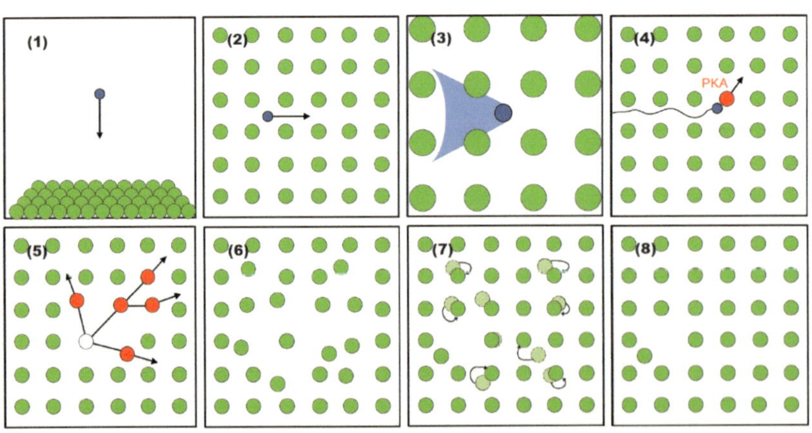

Figure 1.
Schematic representation of different stages of the damage cascade in a crystalline material under irradiation.

electronic defect levels affect the electrical performance of MJ solar cells acting as traps, recombination centers, or carrier removal sites which reduce free carrier concentration [100, 101].

Below, we will present different methods used to describe radiation-induced effects in materials focusing on the description of cumulative effects related to atomic displacements.

4. Monte Carlo particle transport modeling of radiation effects in materials

High-energy charged particles undergo a daunting number of interactions with target materials. Such interactions include:

i. electronic collisions leading to ionization and excitation;

ii. multiple Coulomb scattering at small angles (elastic deflection without energy loss, or minimal inelastic loss);

iii. inelastic nuclear reactions, that is, high-energy reactions in which a nucleus in the target struck by an incident particle (with energy > 10 MeV) undergoes fragmentation into secondary lighter nuclei and other lighter particles;

iv. elastic nuclear interactions (< 10–20 MeV) in which atoms are displaced from their initial positions creating point defects.

The most commonly used approach to study radiation-induced effects in materials is the Monte Carlo (MC) particle transport method [102, 103]. In MC particle transport, the interactions of individual primary ions and their secondaries are sampled to build a history of charged particle passage and energy deposition in the target [104], with a large enough statistical sample of trajectories. The energy- and angle-dependent cross sections for different interactions are provided by theoretical models of the elementary interactions and/or experimental data, depending on the energy window. Codes, such as Geant4 [105], MCNP6 [106–108], FLUKA [109], PHITS [110], and HETC-HEDS [111], have been successfully applied to study the radiation at a hemispherical dome made of lunar regolith used to simulate a lunar habitat [112, 113] and the radiation environment around the Moon [114, 115].

Several relevant radiation-induced effects in materials are due to particles with an energy of a few MeV to a few tenths of MeV, as can be seen in **Table 2**. In this regime, below hadronic interactions causing fragmentation/spallation, atomic displacements are induced in the target by elastic nuclear interactions. Two concepts describe the slowing down of the impacting particles (and the induced secondaries), (i) the *electronic stopping power*, that is, the energy loss of the moving particle to the electronic degrees of freedom of the target (a concept valid in the whole energy range) and (ii) the *nuclear stopping power*, that is, the energy lost to elastic nuclear interactions causing atomic displacements (a concept only used for the regime below hadronic interactions). MC particle transport modeling is a very convenient approach to deal with the enormous amount of interactions that a high-energy particle can induce in a target. However, the approximations used for intermediate and low energies (few MeV and lower) may pose some challenges for the applicability of MC particle transport. At such energies, the atomic-scale structure and the electronic properties of the target system should be taken into account for a reliable description of the radiation-induced effects. In MC particle transport, however, the target materials are amorphous and the macroscopic interaction cross sections, as well as electronic and nuclear stopping power, are obtained by a simple stoichiometric averaging of the elemental cross sections and stopping power. The electronic stopping in MC particle transport simulations is calculated for a uniform electron gas with the same density as the target within the perturbative linear approach not appropriate at low energies [116]. Since the crystal structure of the target is ignored, the effects due to ion channeling (i.e., when the ion path is confined within the crystallographic planes), that have been shown to significantly influence the electronic stopping power [117, 118], are not taken into account.

A displacement cascade in MC particle transport simulations is generally modeled within the Binary Collision Approximation (BCA) [119] which assumes a series of independent two-body collisions. Between collisions, particles travel in a straight line. The BCA is valid when (i) the projectile energy is higher than 1 keV per nucleon, which, for PKAs, could be relevant energy, and (ii) the target material has low density, in which case the collisions between the incoming particle and the target atoms occur rarely. BCA allows reducing the computational complexity of the ion-matter interactions compared to a full many-body simulation (e.g., molecular dynamics, discussed in Section 5) and allows for reaching large dimensions with reduced computational needs. However, this method is valid for linear collisions only and describes only primary damage, that is, it does not account for the dynamic evolution of induced defects at later times (**Figure 2**).

One of the most popular tools in which the BCA is implemented is the Stopping and Range of Ions in Matter (SRIM) code [120]. Besides containing semiempirical data for the electronic stopping power of a variety of targets, SRIM can be applied to model the linear cascades and estimate the number of defects in any material and any ion energy up to 1 GeV. Nuclear stopping in very low-energy intervals uses the so-called ZBL (Ziegler-Biersack-Littmark) universal potential that combines classical Coulomb potential with a semiempirical screening function [120]. The electronic and nuclear degrees of freedom are completely separated in SRIM as well as in other MC particle transport tools used by the particle physics community and the space radiation effects community. Finally, it is important to remark that materials are static in MC particle transport methods—there is no dynamics induced in them by the impact of primaries and the generation and passage of secondaries. Thus, more accurate methods are needed to get access to the processes missing in MC particle transport calculations. Such methods are described in the next section.

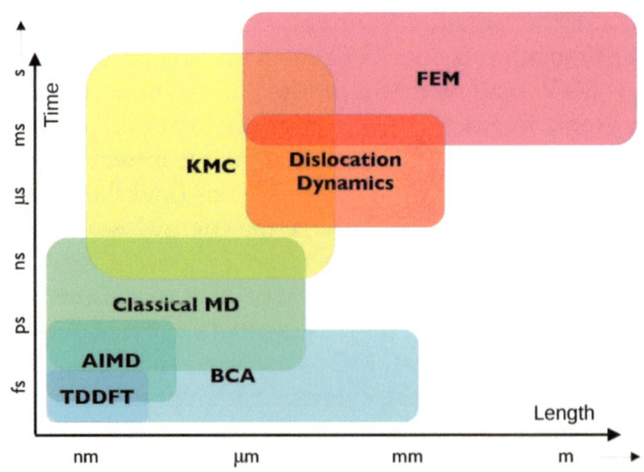

Figure 2.
Time and length scales and corresponding methods can be applied to study different stages of radiation damage.

5. Multiscale approach to modeling radiation damage in materials

There is a large variety of methods used in condensed matter physics and materials science to study radiation effects in materials, each of them describing a particular aspect of the damage process. **Figure 2** shows a schematic representation of the different time and length scales with the corresponding computational methods that can be applied to study different stages of radiation damage [97, 121, 122]. The very first stage, at the smallest time-length scale, is the electronic stopping regime. For decades, the semiempirical SRIM code discussed in the previous section has been the most widely used tool to calculate electronic stopping power. Nowadays, the electronic stopping power (and the induced electronic excitations in the target) can be described by *ab initio* (parameter-free) methods relying on a realistic description of the electronic and ionic properties of the target system. One of the most accurate methods for treating electronic excitations in materials is time-dependent density functional theory (TDDFT) [123] which allows accessing the electronic effects accompanying the ion dynamics. *Ab initio* molecular dynamics (AIMD) [124] based on density functional theory (DFT) [125, 126] can be applied to study point defect formation. At longer time scales and larger length scales, when atomic displacements start to dominate, classical molecular dynamics (MD) and the BCA are usually applied to perform collision cascade simulations. The kinetic Monte Carlo (KMC) [127], the dislocation dynamics (DD) [128], and the finite element method (FEM) [129] are used to study the evolution of defects at the even longer time and larger length scales.

For a complete and accurate description of every aspect of radiation damage, as well as the interplay between them, one has to adopt a combined approach. In recent years, researchers have realized the importance of a multiscale approach to studying radiation damage, as follows from many publications and reviews [121, 122, 130–134]. Each of the methods presented in **Figure 2**, as well the ways of combining them, will be discussed below in the order of increasing complexity. The main focus will be on classical MD, AIMD, and TDDFT, which are fundamental for the description of primary radiation damage at the atomic scale.

5.1 Classical molecular dynamics and beyond: the collision cascade

The most widely used approach in materials science to study the interaction of ions with matter (collision cascades) is MD [135]. MD offers a picture of the ion–ion interaction beyond the linear cascade of the pure BCA by including many-body effects. In MD, atoms are treated as classical particles, and their motion is described by Newtonian dynamics. No electronic effects are thus included.

Cascade simulations need large samples consisting of up to a million atoms (depending on the PKA's energy), which prohibits using parameter-free methods (such as DFT, see Section 5.2) to compute the interatomic forces. Instead, in MD, the forces on atoms are calculated from empirical or semiempirical interatomic potentials (also called force fields) [136–138]. MD with empirical potentials proved to work well for large systems and long time scales [139].

In an MD cascade simulation, the system is usually modeled using periodic boundary conditions, that is, by replicating a small unit cell in all directions. Typically, prior to the cascade simulation itself, a regular MD simulation is done to thermally equilibrate the target system at the desired initial temperature. Then, with the equilibrated configuration, the cascade simulation is initiated by changing the velocity of one of the atoms (the PKA), giving it the desired amount of kinetic energy in the intended direction. The system is then evolved in time as in regular MD, that is, by integrating Newton's equations along with a series of time-steps, which involves computing the atomic forces, velocities, and positions at each time-step (see Refs. [140, 141] for classical texts on MD). At the end of the cascade simulation, the number of defects is obtained by evaluating the final geometry of the system. Usually, cascade simulations are repeated several times, choosing a different PKA and/or a different direction of the PKA's movement to obtain a statistical average of the number of final defects.

MD has been successfully applied to simulate radiation cascades in a variety of materials [139], from simple metals [142, 143] and compounds [144–146] to complex nanostructures [147], 2D materials [148], and novel multicomponent alloys [149, 150]. MD simulations can afford to access the processes taking place on a relatively long time scale up to ps or even ns which is enough to describe the damage cascade until the thermal spike of the collision has dissipated. Most of the MD codes, however, describe only elastic collisions between atoms and disregard the energy loss mechanisms such as electronic excitation and ionization. The possibility of including electronic excitations is discussed in Section 5.3.

After the primary damage has been formed, defects may continue diffusing, thus annihilating or forming defect clusters. Such processes occur on a much longer time scale, reaching at least seconds, not accessible via regular MD. The problem of simulating a process not accessible in a feasible amount of computational time has motivated the development of several enhanced sampling techniques [151], which in the case of MD simulations of materials have allowed to observe otherwise challenging processes, such as phase transitions.

KMC [127] simulations are commonly used to access long-time effects of radiation in materials [152–155]. KMC is designed to model the time evolution of an atomic system. However, instead of solving the equations of motion, as it is done in MD, the KMC method is based on the assumption that the long-time dynamics of a system consists of diffusive jumps from state to state. Each of the states is treated independently, which makes KMC a very efficient method. The dynamics of the system, that is, the probability of transition from one state to another does not depend on the history of the system. The probability of a state-to-state transition is assigned randomly and the most probable transition is statistically chosen. This allows avoiding the complications related to the choice of interatomic potentials, thus overcoming the time limitations of MD simulations (usually $t < 1$ μs) and accessing the macroscopic length scale. KMC is used as an extension of MD to further evolve the damage cascade in time and study the diffusion, accumulation, and annihilation of defects after a collision cascade took place [155].

To further extend the problem into the macro-domain, the DD [128] and FEM [129, 156] methods, based on dividing a geometrical space on a number of finite (non-overlapping) segments, are usually applied. FEM has been used to study the response of a macro-object to external stress in engineering and has also been applied to study the behavior of solids under irradiation by extrapolating the known displacements and evaluating the geometry of a 3D object. DD method allows for calculating the motion of dislocations as well as evaluating the plastic deformation in the material induced by the collective motion of dislocations.

5.2 *Ab initio* molecular dynamics: coupling MD with density functional theory

AIMD is one of the most important tools in quantum physics and chemistry [157]. In a typical AIMD simulation, it is assumed that the system consists of N nuclei and N_e electrons for which the Born-Oppenheimer (BO) approximation is applied [158]. The BO approximation implies that the dynamics of the electronic and nuclear subsystems can be treated separately given the fact that the nuclei are much heavier than the electrons and thus the time scales of their motion are very different. In AIMD, the nuclei are evolved using classical mechanics, while the electronic ground state is adapted to the instantaneous nuclear positions at each step of the dynamics (i.e., the *adiabatic* approximation). The ground-state electronic problem is taken into account through advanced methods, most commonly from DFT [125, 159], a quantum-mechanical method that is used to calculate the electronic structure of many-body systems. Given its importance in describing physical and physicochemical properties of materials, and as it is functional to the understanding of the next section, a brief introduction to the main concepts of DFT is given here.

Practical DFT calculations are based on the Kohn-Sham (KS) formalism [126], which replaces the complex problem of interacting electrons in the standard Schrödinger equation by a problem of non-interacting electrons moving in an effective potential V_{eff}:

$$\left\{-\frac{1}{2}\nabla^2 + V_{\text{eff}}([n], \mathbf{r})\right\}\psi_i^{\text{KS}}(\mathbf{r}) = \varepsilon_i \psi_i^{\text{KS}}(\mathbf{r}), \tag{1}$$

where ε_i is the eigenvalues of the KS equations and ψ_i^{KS} is the one-electron KS wave functions. Here, the effective potential $V_{\text{eff}}([n], \mathbf{r}) = V_{\text{ext}}(\mathbf{r}) + V_{\text{H}}([n], \mathbf{r}) + V_{\text{xc}}([n], \mathbf{r})$ includes the external potential $V_{\text{ext}}(\mathbf{r})$ in which the electrons move (i.e., the electron-nuclei Coulomb attraction), the exchange-correlation (XC) potential $V_{\text{xc}}([n], \mathbf{r})$, in which all the many-body effects are included, and the Hartree potential $V_{\text{H}}([n], \mathbf{r})$ which is the electrostatic potential created by the electron density. The solution to self-consistent KS equations Eq. (1), is the exact electron density of the system of interacting electrons, provided that V_{eff} is known exactly: $n(\mathbf{r}) = \sum_{i=1}^{N} \left|\psi_i^{\text{KS}}(\mathbf{r})\right|^2$. All the properties of the system (e.g., electronic structure and ground-state energy) can be determined from the electron density, according to the Hohenberg-Kohn theorem [125].

AIMD is used to simulate any physicochemical process where the electronic structure of the system changes significantly or when a detailed description of the structure is needed. A typical example would be the simulation of chemical reactions, where chemical bonds are formed or broken, which cannot be described via classical force fields.

5.3 Time-dependent density functional theory for electron dynamics and its coupling to MD

Although the adiabatic BO approximation is the usual approximation in the methods described above, its applicability is only justified in near-equilibrium situations. However, under ion impact, the electronic subsystem is rapidly driven out of equilibrium.

A realistic description of the dynamics of the electrons in the target during the passage of fast ions can be obtained in the framework of TDDFT which gives access to the electron dynamics out of the electronic ground state. In particular, real-time TDDFT [160] provides a non-perturbative description of the electronic excitations upon an external perturbation and can be combined with the Ehrenfest MD scheme [161], which allows for coupling between electron and ion motion, contrary to the BO picture.

TDDFT consists in solving the time-dependent KS equations [123]:

$$i\hbar\frac{\partial}{\partial t}\psi_i^{\text{KS}}(\mathbf{r}, t) = \left\{-\frac{\hbar^2\nabla^2}{2m} + \hat{V}_{ext}(\{R_I(t)\}) + \hat{V}_{HXC}[n(\mathbf{r}, t)]\right\}\psi_i^{\text{KS}}(\mathbf{r}, t), \tag{2}$$

where $\hat{V}_{HXC}$ describes both the electrostatic (Hartree) electron-electron interaction and the quantum-mechanical XC potential, $\hat{V}_{ext}$ is the potential arising from the ions (both the fast-moving impacting particle and the target atoms), and $\{R_I(t)\}$ are the atomic positions. The force on the nuclei in Ehrenfest dynamics is defined as

$$F_I(t) = -\sum_i \langle \psi_i^{\mathrm{KS}}(t) | \nabla_{R_I} \hat{H}_e | \psi_i^{\mathrm{KS}}(t) \rangle \tag{3}$$

where $\hat{H}_e$ is the Hamiltonian, that is, the operator on the r.h.s. of Eq. (2). Ehrenfest dynamics is a mean-field method, meaning that the nuclei move on an effective potential energy surface (a mathematical function that describes the energy of the system and whose value depends on the coordinates of all the atoms), which is an average of all adiabatic states involved, weighted by their populations. Other methods exist, in particular, one allowing for electronic transitions, that is, switches between adiabatic states when their population changes [162].

The solution of the time-dependent KS equations in real time can be obtained by applying the so-called time-evolution operator, evolving the KS states in time [123]. The time-step of this propagation must be of the order of attoseconds to describe the fast dynamics of the electrons, in contrast to what occurs in AIMD and MD where the time-step is of the order of femtoseconds. The time-dependent electron density is calculated at each step, from which the total energy of the system is obtained. Knowing the total energy as a function of time, the electronic stopping power can be calculated as $S_e = -dE/dx$, where dE is the energy loss and dx is the distance traveled by the projectile inside the target.

Many examples of accurate first-principles calculations of the electronic stopping power are available in the literature [117, 118, 163–168]. Recent studies have demonstrated that electronic excitations (induced by both the primary impacting ion and especially by PKAs and further displaced atoms) affect the cascade evolution [118, 169–171] and thus, they need to be accounted for. The electronic stopping effects can be included in MD cascade simulations through the so-called two-temperature (2T) model [118, 172]. In 2 T-MD, the electrons are included as a thermal bath. Each particle is subject to a friction force representing the electronic stopping and a stochastic force representing the coupling between the vibrational degrees of freedom of the lattice and the electrons. This model considers constant electronic density in the entire system and thus, the electronic stopping power is independent of the crystal direction. Recent studies have extended the 2T model by coupling the electronic and nuclear effects via many-body forces that act in a correlated way. This allowed for the construction of a unified model for ion-electron interactions [170, 171, 173, 174] with a complex energy-exchange process between the ionic and electronic subsystems [174].

6. Selected cases of radiation damage studies in materials relevant for exploration of the Moon

The previous section provided an overview of computational methods that can be applied to study radiation damage in materials and discussed the advantages of combining such methods into a multiscale approach. This section mainly focuses on the effects of radiation on materials of practical use on the Moon, including several novel and promising materials. We overview the existing radiation damage studies for these novel materials, emphasizing multiscale modeling when available.

6.1 Improving the model for solar cell degradation via a multiscale approach

Generally, degradation of solar cells is modeled via the non-ionizing energy loss (NIEL) approach, the NIEL being the portion of energy loss per unit path length of the projectile converted into displacement damage. According to Akkerman et al. [175] (the definition used in most simulation tools), the NIEL is defined as:

$$\mathrm{NIEL}(E) = \frac{N_A}{A}\int_{T_d}^{E_{\max}} Q(E_R)E_R\left(\frac{d\sigma}{dE_R}\right)_E dE_R = \frac{N_A}{A}D(E), \tag{4}$$

where E is the total kinetic energy of the external particle, E_R is the portion of kinetic energy which turns into displacement damage, $Q(E_R)$ is the partition factor giving the fraction of kinetic energy to be lost to NIEL mechanisms, N_A is Avogadro's number, A is the atomic mass of the lattice atom, $d\sigma/dE_R$ is the partial differential cross section for creating a recoil atom with energy E_R, and $D(E)$ is the displacement damage function. The integral runs from the minimum energy required to permanently displace an atom to a defect position, that is, the threshold displacement energy T_d, to $E_{\max}$, which is the maximum energy transferred to a recoil atom in a particular interaction. Although the NIEL concept differs from the nuclear stopping power, as it includes also the energy loss to non-ionizing events induced by hadronic interactions, as already mentioned above (see **Table 2**), relevant non-ionizing effects are induced by particles with energies from few to few tenths of MeV, which is the regime of Coulomb interactions.

On the basis of a large set of experimental observations, it is assumed that the degradation of a semiconductor device under irradiation can be linearly correlated with the NIEL [176]. In practice, this means that the number of defects should give a measure of the damage irrespective of their distribution, whether clustered in high density in small regions (as in the case of neutron damage) or homogeneously scattered over a relatively wide volume (as in the case of the low-energy proton or γ-ray-induced damage) [177]. Thus, in principle, the damage produced by different particles (with different energies) should be scalable via their NIEL (i.e., the number of displacements), as indeed has been shown in several studies [176, 178–182]. The NIEL scaling is a powerful method for dealing with displacement damage predictions in complex radiation environments, such as on the Moon and in space missions in general. However, deviations from the linearity exist and seem to be associated with the "quality" of the radiation damage at the microscale as induced by different kinds of particles and (or) as influenced by intrinsic defects in the target [183].

Generally, the NIEL is calculated via MC particle transport codes, assuming amorphous target materials, a static T_d that is constant for each element in all the materials where such element is found (thus, not considering the underlying electronic structure), and a simple linear collision cascade model for the number of final defects [184]. Several quantities in the NIEL formula and, more generally, the overall understanding of radiation damage can be strongly improved via AIMD and TDDFT+MD studies. T_d, for example, is an important quantity that can significantly affect the NIEL [185, 186] and its accurate estimation can be accessed by AIMD simulations [187, 188]. Recent results for T_d in semiconductors have shown that the electronic excitations can, in general, reduce their value [122], in line with previous experimental results [189]. The effect of electronic excitations consists in weakening the atomic bonds making it easier to displace an atom from its equilibrium position. On the other hand, the "heating" effect of electronic excitations has the consequence of facilitating the healing of the structure. Even a small change in T_d affects the calculated NIEL, as can be seen in **Figure 3** showing the NIEL for a

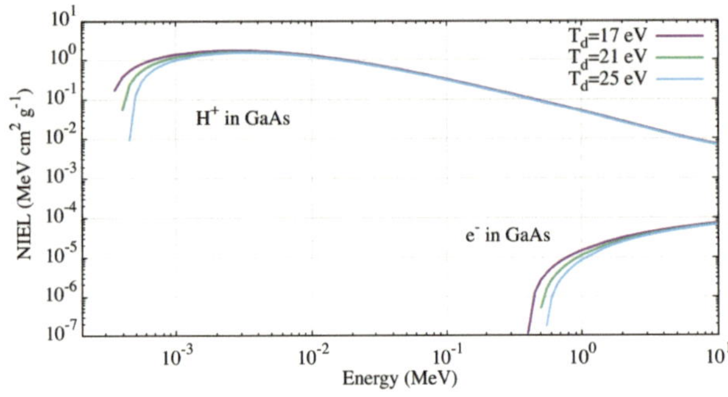

Figure 3.
NIEL for protons and electrons in GaAs for different values of the threshold displacement energy T_d calculated with the online SR-NIEL tool [190].

proton and an electron in GaAs. The NIEL is affected by the choice of T_d in an energy window of the lunar radiation environment (see **Table 1**). These findings raise an important question about the role of electronic excitations in defect formation that deserves more attention in future works.

Another example of possible improvement in the NIEL model is a more precise calculation of the number of radiation-induced defects and of the “quality” of radiation-induced damage (which type of defects are induced). It has been observed that point-like and clustered defects contribute differently to some degradation parameters [191]. Recent MD studies [192–194] and experimental works [181, 195, 196] have proposed an effective or *adjusted* NIEL to correct the deviations from a linear dependence of degradation parameters on the NIEL and restore a linear relationship. Other MD studies [197] proposed new metrics for counting defects including the effect of a “heat spike”, which leads to a much lower rate of final defects as compared to predictions from a simple linear collision cascade model as commonly used in the NIEL calculations based on MC particle transport [184, 198].

On a parallel research stream, multiscale studies in a number of materials combining MD simulations of collision cascades with the electronic stopping from TDDFT offer a more accurate description of both the number and the nature of defects created under realistic conditions. The electronic degrees of freedom and their coupling to the phonons of the target affect the cascade evolution and morphology [170, 171, 173, 174]. This is of relevance for the NIEL which includes a part of energy dissipated to phonons. This fraction depends on the energy of the impinging particle but also on the properties of the material. Some studies have shown that the direction-dependence of the electronic stopping can influence the collision cascades [118]. Other studies have demonstrated that the formation of thermal spikes and therefore of amorphous pockets is sensitive to the electronic specific heat [199] and others that the choice of the model employed for the inclusion of the electronic effects and in particular the overestimation (or underestimation) of electron-phonon coupling can have a significant influence on the number of defects created [171].

6.2 Radiation effects in the next-generation lightweight photovoltaic panels

As discussed in Section 2, HOIPs have a unique combination of properties particularly interesting for lunar exploration. The general chemical formula for perovskites is ABX_3, where A and B are two metal ions with different ionic radii and X is an anion that is six coordinated to the B-site [200]. HOIPs, in particular,

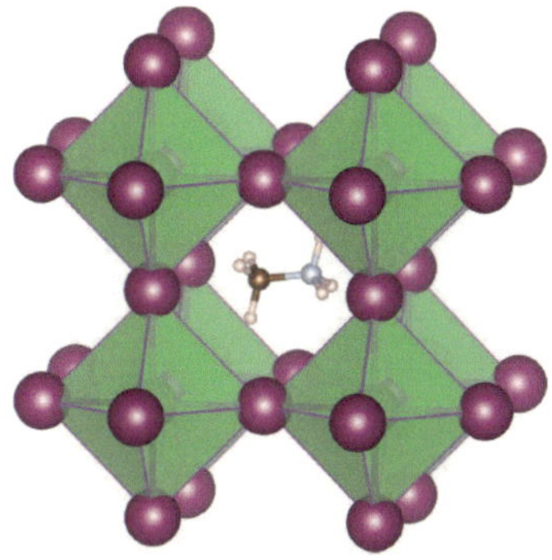

Figure 4.
Structure of a HOIP: methylammonium cation ($CH_3NH_3^+$) occupies the central A site surrounded by 12 nearest-neighbor iodide ions in corner-sharing PbI_6 octahedra [201] (available under the terms of the creative commons CC BY license).

comprise a negatively charged lead-halide inorganic skeleton where B is a metal cation (Sn^{2+} or Pb^{2+}), X is a halide anion (I^-, Br^-, and/or Cl^-) and A is a monovalent positively charged organic cation, such as methylammonium ($MA^+ = CH_3NH_3X^+$, where X = I, Br, Cl) or formamidinium ($FA^+ = CH(NH_2)_2^+$ (**Figure 4**).

Despite many advantages, several external factors, such as air, moisture [202], UV light [47, 203], heat, light soaking [204], and partially also radiation [205, 206], induce considerable structural instabilities in HOIPs. An intrinsic instability is also present, caused by a relatively weak cohesion between the organic cation and the inorganic octahedra and predominantly by the low-energy barriers for the migration of halide anions and organic cations, with halide migration being the most prevalent [201, 207–210]. Phase segregation can be induced by large-scale ion migration [211]. However, some of the challenges that HOIPs-based solar cells face on Earth, such as degradation caused by moisture, are not relevant for space applications [212]. Thermal and vacuum stability, high power-conversion efficiency, and radiation resistance are the main challenges in the space context. A sensible choice of the chemical composition, of eventual use in tandem devices [212] (which also helps to reach an efficiency of up to 30%) or incorporation of a functionalized 2D metal-organic frameworks (MOFs) [213], can improve the long-term operational stability of HOIPs.

A relevant collection of DFT studies for HOIPs can be found in Ref. [214]. A recent study based on DFT + compressed sensing-symbolic regression has shown that mitigation of the propensity of halogens to migrate could be achieved by selectively strengthening specific bonds [215]. The study also unveiled the reasons for improved stability given by specific halogens, the origin of the higher stability offered by certain organic cations compared to others, and highlighted in a quantitative and first-principles manner how weak interactions have a significant role in binding the halogens more strongly.

The study of the radiation tolerance of perovskite solar cells is an extremely active field of research. Solar cells based on HOIPs as active layers have been recently sent to space via first campaigns [60, 216]. Several ground-testing experiments have been performed mostly using protons, either with an energy of several tenths of MeV [69, 211, 217] or with an energy of 150 keV, 100 keV, and 50 keV [70, 218, 219], of less relevance for realistic space conditions.

Superior radiation resistance of perovskite solar cells in comparison to commercially available crystalline Si-based cells has been demonstrated [69]. Moreover, experiments have shown that perovskite solar cells have remarkable self-healing

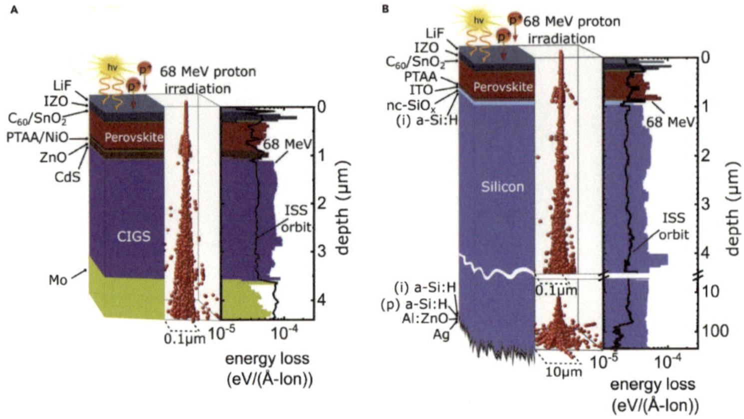

Figure 5.
3D scatter plots of the straggling of 68 MeV protons within the (A) HOIP/CIGS(Cu(In,Ga)Se_2) and (B) HOIP/SHJ(Si heterojunction) tandem solar cells. The energy loss of the incident 68 MeV protons to recoils is plotted as a function of depth based on SRIM simulations with a total of 5×10^7 protons. The damage of a real space environment at the orbit of the ISS is shown as a black line. Adapted from [90] (available under the terms of the Creative Commons CC-BY license).

capabilities (at room temperature) that lower the number of defects caused by proton irradiation [69]. Another experimental study has shown that the proton irradiation effects on the physical properties of HOIPs are strongly dependent on the synthesis method [220] which appeared to affect the strength of specific chemical bonds. In particular, HOIPs, produced by mechano-chemical synthesis, have shown practically no change in their physical properties after irradiation with a high-energy 10 MeV proton beam with doses of up to 10^{13} protons/cm^2.

Recently, multi-junction tandem solar cells (combining HOIPs with previous technologies or technologies investigated in parallel) have also been studied under ion irradiation [217]. Lang et al. [217] carried out SRIM simulations of energy loss of high-energy protons as well as the energy transferred to the recoiling nuclei—a measure of the degradation of PV parameters—in tandem solar cells (**Figure 5**). The study [217] has shown that HOIP/CIGS tandem solar cells possess a high radiation hardness and retain over 85% of their initial performance even after 68 MeV proton irradiation and a dose of 2×10^{12} proton/cm^2, equivalent to 50 years in space at the International Space Station (ISS) orbit.

First-principles calculations of the atomic knock-on displacement events in HOIPs have shown that such displacements are significant and highly energy-dependent [221]. The work has shown that only certain types of atoms are prone to displacements suggesting that mitigation strategies should be directed toward some chemical species more than others. Overall, further studies are necessary, but existing research proves that HOIPs-based solar cells have a remarkable potential for power generation on missions to low Earth orbit, the Moon, and beyond [62].

6.3 Novel multi-principal-element alloys with enhanced radiation resistance

6.3.1 Outstanding properties of MPEA for space applications

Another promising class of novel materials for space applications is multi-principal element alloys (MPEAs) [222, 223], which combine superior mechanical properties and enhanced radiation resistance [224]. Also known as high-entropy alloys (HEAs) or concentrated solid-solution alloys (CSSAs), MPEAs consist of at least five principal elements with the concentration of each element from 5 to 35%

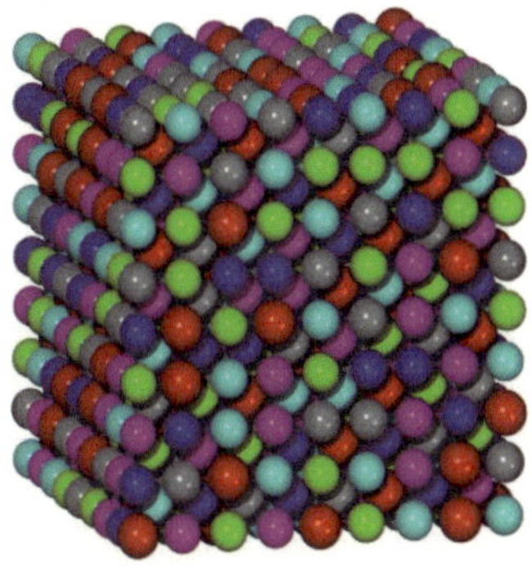

Figure 6.
Atomic structure of a body-centered cubic (BCC) AlCoCrCuFeNi HEA. The Al, Fe, Co, Cr, Ni, and Cu atoms are shown in red, magenta, green, blue, cyan, and gray colors, respectively [225] (available under the https://creativecommons.org/licenses/by-nc-sa/3.0/Creative Commons Attribution License).

[222]. Despite the complex composition, MPEAs often form single-phase solid solutions (**Figure 6**). The interest of researchers in MPEAs has been growing exponentially in recent years, as they exhibit a paradigm shift in alloy development. MPEAs indeed combine a set of outstanding properties, such as high strength, hardness, fracture toughness, corrosion resistance, strength retention at high temperature [226], good low-temperature performance [227], and recently discovered enhanced radiation resistance, superior to conventional alloys and pure metals [149, 222, 223, 228–233]. Moreover, MPEAs have great potential as 3D printing materials [234]. MPEAs can be printed from a powder, providing manufacturing freedom for lightweight and customizable products of complex geometries for applications in the aerospace, energy, molding, tooling, and other industries, all of the great relevance for the exploration of the Moon.

Recent experiments have shown that MPEAs have a higher resistance to defect formation due to high atomic-level stress and chemical heterogeneity [235]. MPEAs also possess lower void swelling and higher phase stability [236, 237] as compared to conventional alloys. Self-healing capability is another remarkable property of MPEAs [227, 236, 238].

The subclass of lightweight (LW) MPEAs have a great potential for space applications due to their high strength-to-weight ratio [239–241]. The main components of LWMPEAs are low-density elements, such as Al, Mg, Si, and Ti [240]. The latter is of extreme importance for ISRU since 99% of the lunar soil consists of Si, Al, Ca, Fe, Mg, and Ti oxides [5, 242].

Currently, the main focus of computational studies has been on the single-phase random solid-solution (SS) alloys based on transition metals with high densities (Co, Cr, Fe, Ni) for application in radiation environments, in particular in nuclear reactors [148, 149, 232, 236, 243–245]. MD simulations of displacement cascades applied to pure metals and multicomponent alloys [150, 244–248] confirm the experimentally observed reduction of the number of defects and defect clusters in MPEAs compared to pure metals (**Figure 7**).

The electronic stopping power for a proton in binary alloys has recently been calculated using real-time TDDFT [249]. The study has shown that the electronic stopping power of binary alloys is higher than that of pure Ni, suggesting that alloys more effectively stop the incoming particles. Moreover, the inclusion of the electronic stopping into MD simulations of defect formation significantly reduces the final number of surviving defects, as shown in **Figure 8**. The inclusion of both the electron-phonon coupling and the electronic stopping in the 2T-MD model not only reduces the actual number of defects but also notably impacts their final

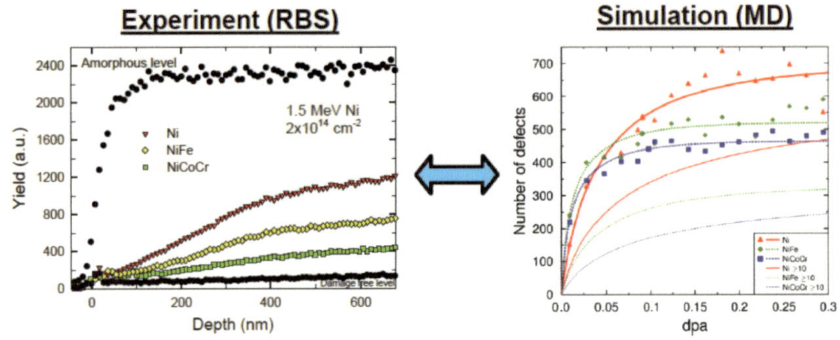

Figure 7.
The number of defects in Ni, NiFe, and NiCoCr from experiments and MD simulations [150].

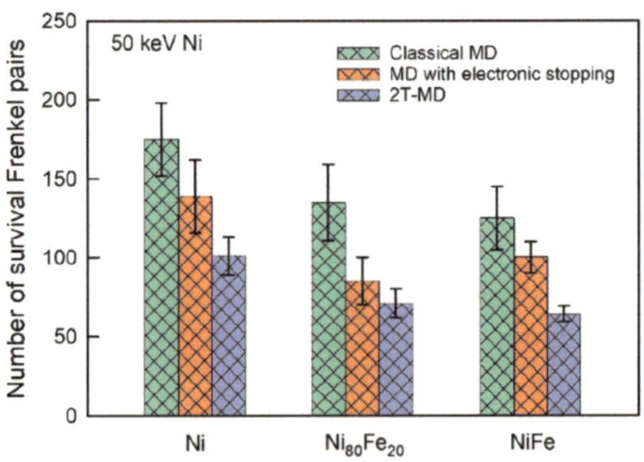

Figure 8.
Average number of surviving defects in the classical MD cascade, MD cascade including electronic stopping force, and the 2 T-MD cascade at the end of the simulation for 50 keV Ni cascade in Ni, $Ni_{80}Fe_{20}$, and NiFe [224, 250–252].

arrangement, namely leading to more isolated point defects and reducing the size of defect clusters in binary and ternary alloys [250–254].

The majority of MD studies focus on binary and ternary MPEAs due to the lack of force fields for alloys with more than three elements. However, some studies exist [233] on defect formation in NiCoFeCr alloy in which fewer defects have been found at the end of the displacement cascade with PKA energies from 10 to 50 keV, as compared with pure Ni. The limitations of the classical MD with force fields and the ways of solving this problem are discussed in the following.

6.3.2 Machine-learning assisted materials discovery

Classical MD with empirical potentials is the method that proved to work well for large systems and long time scales [139] for the modeling of collision cascades. However, classical interatomic potentials cannot accurately reproduce interactions between the atoms in MPEAs due to their complex structure and lattice distortions leading to internal strain [149, 255, 256]. On the other hand, *ab initio* methods relying on quantum mechanics, such as DFT, can accurately reproduce the interatomic potentials in complex structures but are limited to a small length scale.

Recent developments in machine learning (ML) approaches can provide a solution to this problem. ML-enhanced materials discovery is an emerging and extremely rapidly growing field. The combination of a precise model based on quantum mechanics and ML algorithms have the potential for an efficient and accurate description of materials properties [257–259]. Much progress has been made in recent years in the development of ML-based interatomic potentials with the input from electronic structure calculations. First applications have shown that accurate potentials can be obtained for many relevant systems [260–265]. ML-assisted calculations have been applied to pure metals, binary, ternary alloys [266, 267], and MPEAs [268–270].

ML and artificial intelligence (AI) may become powerful tools for more accurate multiscale modeling of materials properties. Artificial Neural Networks (ANN) [271] combined with atomistic KMC have already been used to describe the microstructural changes in metals and alloys induced by irradiation [272]. Machine-learned interatomic potentials have been used to study defect formation in refractory MPEAs [273]. The results confirm experimental findings, showing that the 3D migration and increased mobility of defects in MPEAs promote defect recombination leading to more efficient healing. AI, thus, can provide a bridge between different methods, such as DFT, MD, and KMC, and allow for large-scale atomistic simulations of high accuracy, which will accelerate the discovery of new advanced materials.

6.4 Radiation resistance of fiber-reinforced polymers and composites for habitats

Fiber-reinforced polymers (FRPs) are composite materials made of a polymer matrix reinforced with fibers. Typical polymers that are often used include epoxy, vinyl ester, polyester thermosetting plastic, and phenol-formaldehyde resins. Typical fibers include, but are not limited to, glass, carbon, and aramid. In a composite FRP material, the polymer and fiber often have significantly different physical and/or chemical properties, which remain separate and distinct within the finished structure but are complementary for tailored properties [274]. Because of their low density (lightweight), great moldability, specific strength, stiffness [275], excellent mechanical stability, and good thermal properties, FRPs are being increasingly used as structural materials in aerospace, automotive, marine industries, and civil infrastructures. Hence, FRPs are of great interest for many applications for lunar missions as potential structural materials [276]. Glass fibers (also "fiberglass") can be directly produced from the lunar soil as well as from by-products of metal extraction and can be used to reinforce lunar concrete [277].

The radiation environment on the Moon presents challenges for FRPs with concerns on both the immediate reactions taking place in the materials (short-term effects) and continued post-exposure degradation processes (long-term effects) [276, 278]. In the past decades, many selected FRPs have been ground-tested at different kinds of radiation and particle accelerator facilities for their potential use in space-related radiation environments, including UV-light [279, 280], γ-rays [281, 282], electron beams [283, 284] and proton beams [285].

Carbon-fiber composites have been widely used in aerospace industries due to their high-temperature stability and low density along with high strength, as well as superior beam-induced shock absorption [285, 286]. A combined modeling and experimental study of the radiation effect on carbon-fiber-reinforced molybdenum-graphite compound (MoGRCF) [285], including MC simulations of the energy deposited into a realistic structure by a 200-MeV proton beam (**Figure 9**) has show that carbon-fiber-reinforced composites have superior beam-induced shock absorption ability compared to that of graphite.

In the 1980s, the degradation behavior of carbon-fiber-reinforced plastic (CFRP) under electron beam irradiation in various conditions simulating experiments in space has been studied by Sonoda et al. [283]. It has been observed that

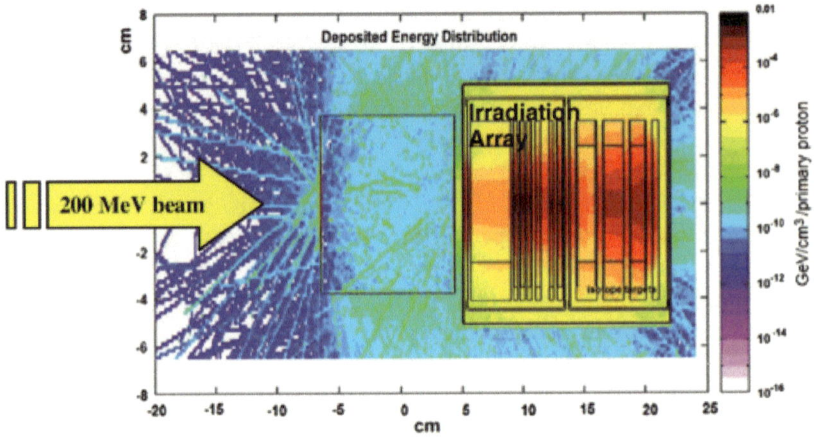

Figure 9.
MC modeling of the energy deposition for a 200-MeV proton beam interacting with an irradiation target array (MoGRCF) in tandem with the isotope production array downstream [285] (available under the terms of the Creative Commons Attribution 3.0 License).

there is no change in mechanical properties of CFRP when irradiated by up to a dose of 50 MGy. MC simulations of radiation effects in FRPs have shown that by adding lead nanoparticles it is possible to increase their radiation resistance [287]. According to the study, the addition of 15 wt% of lead nanoparticles to FRPs led to a mass reduction of ~64% for the same level of radiation shielding.

An alternative to glass fiber for polymer reinforcement is basalt fiber which offers advantages, such as high specific mechanical and physicochemical properties, biodegradability, non-abrasive qualities, and cost-effectiveness [288]. Arnhof et al. [289] have recently studied mechanical properties of fiber-reinforced geopolymer (FRG) with basalt fiber (i.e., inorganic alumino-silicate polymer) made from lunar regolith simulant as potential shielding and structural material. As basalt fibers can be produced *in situ* at the lunar surface efficiently, they can be used widely to increase the mechanical strength of geopolymers. Overall, geopolymers are advantageous for lunar construction due to their excellent resistance to extreme temperature fluctuations and adequate shielding from radiation [290], as well as enhanced mechanical properties over conventional cement [291].

The additive-manufacturing (AM) techniques for lunar construction from regolith, including FRP materials, and their suitability for ISRU has recently been reviewed in Refs. [292, 293]. The AM techniques for lunar construction include Cement Contour Crafting (CCC), Binder Jetting (BJ), Selective Solar Light Sintering (SSLS) and Selective Laser Sintering/Melting (SLS/SLM) for 3D printing and metal melting, Stereolithography/Digital Light Processing (SLA/DLP), among others. CCC and BJ technologies could be used for outdoor lunar civil engineering. SSLS could be applied to both direct compacting of lunar regolith to ceramic parts and 3D printing. SLA/DLP-based methods could be used for the indoor manufacturing of ceramic instruments, providing higher precision and printing quality and lower defect rate of the printed parts than other AM methods. In the last decade, studies have clearly shown that the 3D-printing technologies will become one of the cornerstones of lunar exploration, providing future astronauts with all the necessary infrastructure [293].

Lunar concrete consisting of mined regolith with the addition of glass fibers (also made *in situ* from regolith containing plenty of silicates) has a high strength-to-weight ratio and can be easily 3D-printed, as tests on lunar regolith simulant have shown [294]. Other studies have shown the promising properties of urea from astronauts' urine as a superplasticizer for lunar geopolymers for 3D-printing applications. The use of urea is expected to reduce the necessary amount of water by about 30% [295].

It is worth mentioning that the 4D printing of a "smart material" with FRPs that responds to radiation-induced damages and aging in a programmable way could be realized in near future [296, 297]. In addition to experiments on the radiation environment in a lab, multiscale computational simulations as mentioned above could be helpful for gaining further insights into the radiation-induced molecular changes occurring in polymers.

7. Conclusions

In this chapter, we introduced some relevant materials for lunar habitat construction and power generation. We discussed the radiation environment on the Moon and the effects that radiation can cause in such materials. We provided an overview of computational methods used to study different stages of radiation damage in materials, focusing on the methods that allow simulating the behavior of materials with extreme accuracy down to the atomic scale. We emphasized that by coupling different methods, it is possible to account for different time and length scales in the evolution of the radiation-induced effects and to combine the electronic effects with atomic displacements.

Several particular examples of radiation damage studies have been discussed with the focus on novel materials with enhanced radiation resistance and other remarkable properties for use on the Moon that can revolutionize space exploration. Such materials include HOIPs for energy production and MPEAs and FRP composite materials for construction. The primary materials considered for lunar construction are FRGs with basalt or glass fibers, which have excellent mechanical properties, can benefit from ISRU, and provide necessary radiation shielding. We emphasized that researchers' effort is mainly directed toward the development of additive manufacturing techniques, such as 3D printing for habitat construction from lunar regolith. 3D printing will allow producing complex and customizable products in a shorter time and with a lower cost and material consumption.

Nowadays, the radiation-induced effects in materials for space missions are mainly studied by MC particle transport modeling, inheriting the remarkable modeling and computational efforts by the high-energy physics community. However, with the development of first-principles methods and multiscale simulations, a more accurate understanding of radiation effects in materials can be achieved for the regime below hadronic interactions, with details down to atomic scale. It can be expected that the combination of first-principles methods, MC particle transport, and ML will contribute further to the investigation of materials to unravel their full potential for the application in harsh space radiation environments, in particular for what concerns the resistance and resilience to cumulative displacements effects.

Acknowledgements

The authors are grateful for the funding provided by the project ESC2RAD within the Horizon 2020 Research and Innovation program (grant agreement ID: 776410) and by the project PROIRICE within the program H2020-MSCA-IF 2016 of the Horizon 2020 program of the European Union (grant agreement ID: 748673).

Author details

Natalia E. Koval[1*], Bin Gu[2,3], Daniel Muñoz-Santiburcio[1,4] and Fabiana Da Pieve[5]

1 CIC Nanogune BRTA, San Sebastián, Spain

2 Jiangsu International Joint Laboratory on Meteorological Photonics and Optoelectronic Detection, Nanjing University of Information Science and Technology, Nanjing, China

3 Atomistic Simulation Centre, Queen's University Belfast, Belfast, United Kingdom

4 Instituto de Fusión Nuclear "Guillermo Velarde", Universidad Politécnica de Madrid, Spain

5 Royal Belgian Institute for Space Aeronomy BIRA-IASB, Brussels, Belgium

*Address all correspondence to: natalia.koval.lipina@gmail.com

References

[1] Crawford IA, Anand M, Cockell CS, Falcke H, Green DA, Jaumann R, et al. Back to the Moon: The scientific rationale for resuming lunar surface exploration. Planetary and Space Science. 2012;**74**(1):3-14. Scientific Preparations For Lunar Exploration

[2] Off-Earth Manufacturing Symposium: How to Build a New Home in Space. 2021. Available from: https://www.esa.int/Enabling_Support/Preparing_for_the_Future/Discovery_and_Preparation/Off-Earth_manufacturing_symposium_how_to_build_a_new_home_in_space [Accessed: December 12, 2021]

[3] NASA's Plan for Sustained Lunar Exploration and Development. 2020. Available from: www.nasa.gov/sites/default/files/atoms/files/a_sustained_lunar_presence_nspc_report4220final.pdf [Accessed: December 12, 2021]

[4] NASA's Lunar Exploration Program Overview. 2020. Available from: https://www.nasa.gov/sites/default/files/atoms/files/artemis_plan-20200921.pdf [Accessed: December 12, 2021]

[5] Stefanescu DM, Grugel RN, Curreri PA. In situ resource utilization for processing of metal alloys on Lunar and Mars BASES. In: Space 98. Albuquerque, MN: American Society of Civil Engineers; 1998. pp. 266-274

[6] Zacny K. Lunar drilling, excavation and Mining in Support of science, exploration, construction, and In situ resource utilization (ISRU). In: Badescu V, editor. Moon: Prospective Energy and Material Resources. Berlin Heidelberg, Berlin, Heidelberg: Springer; 2012. pp. 235-265

[7] Just GH, Smith K, Joy KH, Roy MJ. Parametric review of existing regolith excavation techniques for lunar in situ resource utilisation (ISRU) and recommendations for future excavation experiments. Planetary and Space Science. 2020;**180**:104746

[8] Sacksteder K, Sanders G. In-situ resource utilization for Lunar and Mars exploration. AIAA 2007-345. 45th AIAA Aerospace Sciences Meeting and Exhibit; Reno, Nevada. US: NTRS; 2007

[9] Schreiner SS, Sibille L, Dominguez JA, Hoffman JA. A parametric sizing model for molten Regolith electrolysis reactors to produce oxygen on the Moon. Advances in Space Research. 2016;**57**(7):1585-1603

[10] Crawford IA. Lunar resources: A review. Progress in Physical Geography: Earth and Environment. 2015;**39**(2):137-167

[11] Schwandt C, Hamilton JA, Fray DJ, Crawford IA. The production of oxygen and metal from lunar regolith. Planetary and Space Science. 2012;**74**(1):49-56. Scientific Preparations For Lunar Exploration

[12] Kozyrovska N, Lutvynenko TL, Korniichuk OS, Kovalchuk M, Voznyuk TM, Kononuchenko O, et al. Growing pioneer plants for a lunar base. Advances in Space Research. 2006;**37**:93-99

[13] Maggi F, Pallud C. Space agriculture in micro- and hypo-gravity: A comparative study of soil hydraulics and biogeochemistry in a cropping unit on earth, Mars, the Moon and the space station. Planetary and Space Science. 2010;**58**(14):1996-2007

[14] Ghidini T. Materials for space exploration and settlement. Nature Materials. 2018;**17**(10):846-850

[15] Pernigoni L, Lafont U, Grande AM. Self-healing materials for space applications: Overview of present

development and major limitations. CEAS Space Journal. 2021;**13**(3): 341-352

[16] Belvin W, Watson J, Singhal S. Structural concepts and materials for lunar exploration habitats. AIAA-2006-7338. AIAA Space 2006 Conference and Exposition; San Jose, CAL. US: NTRS; 2006

[17] Experimental Investigation of Selective Laser Melting of Lunar Regolith for In-Situ Applications. Volume 2A: Advanced Manufacturing of ASME International Mechanical Engineering Congress and Exposition, 11 2013. V02AT02A008

[18] Benaroya H. Lunar habitats: A brief overview of issues and concepts. REACH. 2017;**7-8**:14-33

[19] Bodiford M, Burks K, Perry M, Cooper R, Fiske M. Lunar in situ materials-based habitat technology development efforts at NASA/MSFC. In: Earth & Space 2006: Proceedings of 10th ASCE Aerospace Division International Conference on Engineering, Construction, and Operation in Challenging Environments. US: ASCE; 2006

[20] Ruess F, Schaenzlin J, Benaroya H. Structural design of a lunar habitat. Journal of Aerospace Engineering. 2006; **19**(3):133-157

[21] Rojdev K. Long term lunar radiation degradation of potential lunar habitat composite materials [PhD thesis]. 2012

[22] Meurisse A, Cazzaniga C, Frost C, Barnes A, Makaya A, Sperl M. Neutron radiation shielding with sintered lunar regolith. Radiation Measurements. 2020;**132**:106247

[23] Akisheva Y, Gourinat Y. Utilisation of Moon Regolith for radiation protection and thermal insulation in permanent lunar habitats. Applied Sciences. 2021;**11**(9):3853

[24] Miller J, Taylor LA, DiGiuseppe M, Heilbronn LH, Sanders G, Zeitlin CJ. Radiation shielding properties of lunar regolith and regolith simulant. LPI Contributions. 2008;**1415**:2028

[25] Montes C, Broussard K, Gongre M, Simicevic N, Mejia J, Tham J, et al. Evaluation of lunar regolith geopolymer binder as a radioactive shielding material for space exploration applications. Advances in Space Research. 2015;**56**(6):1212-1221

[26] Wang KT, Lemougna PN, Tang Q, Li W, Cui XM. Lunar regolith can allow the synthesis of cement materials with near-zero water consumption. Gondwana Research. 2017;**44**:1-6

[27] Toutanji H, Fiske MR, Bodiford MP. Development and Application of Lunar "Concrete" for Habitats. In: Earth & Space 2006: Engineering, Construction, and Operations in Challenging Environment. US: ASCE; 2006

[28] Lim S, Anand M. Space Architecture for Exploration and Settlement on Other Planetary Bodies: In Situ Resource Utilisation (ISRU) Based Structures on the Moon: Proceedings of the 2nd European Lunar Symposium; London, UK. 2014

[29] Meurisse A, Makaya A, Willsch C, Sperl M. Solar 3D printing of lunar regolith. Acta Astronautica. 2018;**152**: 800-810

[30] Cesaretti G, Dini E, De Kestelier X, Colla V, Pambaguian L. Building components for an outpost on the lunar soil by means of a novel 3D printing technology. Acta Astronautica. 2014;**93**: 430-450

[31] Shaw M, Humbert M, Brooks G, Rhamdhani A, Duffy A, Pownceby M. Mineral processing and metal extraction on the lunar surface—challenges and opportunities. Mineral Processing and

Extractive Metallurgy Review. 2021; **0**(0):1-27

[32] Dietzler D. Making it on the Moon: Bootstrapping Lunar industry. In: NSS Space Settlement Journal. Vol. 1. Washington, DC, USA: National Space Society; 2016

[33] Palos MF, Serra P, Fereres S, Stephenson K, González-Cinca R. Lunar isru energy storage and electricity generation. Acta Astronautica. 2020; **170**:412-420

[34] Ellery A. Generating and storing power on the moon using in situ resources. Proceedings of the Institution of Mechanical Engineers, Part G: Journal of Aerospace Engineering. 2021;**0**(0): 1-19

[35] Landis G, Bailey S, Brinker D, Flood D. Photovoltaic power for a lunar base. Acta Astronautica. 1990;**22**: 197-203

[36] Halbach E, Inocente D, Katz N, Petrov G. Solar Arrays with Variable Panel Elevations for the Moon Village. 2021

[37] Abhirama RK. Power generation system for lunar habitat. Technical Report. Technische Universität Berlin; 2020

[38] Kohout LL. Cryogenic Reactant Storage for Lunar Base Regenerative Fuel Cells: Proceedings of the International Conference on Space Power; Cleveland, OH. 1989

[39] Ellery A. Supplementing closed ecological life support systems with in-situ resources on the Moon. Life. 2021;**11** (8):770

[40] Bolonkin A. Inflatable Dome for Moon, Mars, Asteroids, and Satellites. AIAA-2007-6262. AIAA SPACE 2007 Conference & Exposition. Long Beach, California. US: NTRS; 2007

[41] Li J, Aierken A, Liu Y, Zhuang Y, Yang X, Mo JH, et al. A brief review of high efficiency III-V solar cells for space application. Frontiers in Physics. 2021;**8**:657

[42] Spitzer MB, Fan JCC. Multijunction cells for space applications. Solar Cells. 1990;**29**(2):183-203

[43] Torchynska T, Polupan G, Conde Zelocuatecatl F, Scherbina E. Application of iii-v materials in space solar cell engineering. Modern Physics Letters B. 2001;**15**(17-19):593-596

[44] Best Research-cell Efficiency Chart. 2021. Available from: www.nrel.gov/pv/cell-efficiency.html [Accessed: January 2021]

[45] Huang J, Yuan Y, Shao Y, Yan Y. Understanding the physical properties of hybrid perovskites for photovoltaic applications. Nature Reviews Materials. 2017;**2**(7):17042

[46] Schmidt-Mende L, Dyakonov V, Olthof S, Ünlü F, Lê KMT, Mathur S, et al. Roadmap on organic–inorganic hybrid perovskite semiconductors and devices. APL Materials. 2021;**9**(10): 109202

[47] Berry J, Buonassisi T, Egger DA, Hodes G, Kronik L, Loo Y-L, et al. Hybrid organic-inorganic perovskites (HOIPs): Opportunities and challenges. Advanced Materials. 2015;**27**(35): 5102-5112

[48] Wehrenfennig C, Eperon GE, Johnston MB, Snaith HJ, Herz LM. High charge carrier Mobilities and lifetimes in Organolead Trihalide perovskites. Advanced Materials. 2014;**26**(10): 1584-1589

[49] Brenner TM, Egger DA, Kronik L, Hodes G, Cahen D. Hybrid organic— inorganic perovskites: Low-cost semiconductors with intriguing

charge-transport properties. Nature Reviews Materials. 2016;**1**(1):15007

[50] Saba M, Quochi F, Mura A, Bongiovanni G. Excited state properties of hybrid perovskites. Accounts of Chemical Research. 2016;**49**(1):166-173

[51] Motta C, El-Mellouhi F, Sanvito S. Charge carrier mobility in hybrid halide perovskites. Scientific Reports. 2015; **5**(1):12746

[52] Ponseca CS, Savenije TJ, Abdellah M, Zheng K, Yartsev A, Pascher T, et al. Organometal halide perovskite solar cell materials rationalized: Ultrafast charge generation, high and microsecond-long balanced Mobilities, and slow recombination. Journal of the American Chemical Society. 2014;**136**(14): 5189-5192

[53] Stranks SD, Eperon GE, Grancini G, Menelaou C, Alcocer MJP, Leijtens T, et al. Electron-hole diffusion lengths exceeding 1 micrometer in an organometal trihalide perovskite absorber. Science. 2013;**342**(6156): 341-344

[54] Kojima A, Teshima K, Shirai Y, Miyasaka T. Organometal halide perovskites as visible-light sensitizers for photovoltaic cells. Journal of the American Chemical Society. 2009; **131**(17):6050-6051

[55] Yue L, Yan B, Attridge M, Wang Z. Light absorption in perovskite solar cell: Fundamentals and plasmonic enhancement of infrared band absorption. Solar Energy. 2016;**124**:143-152

[56] Snaith HJ. Perovskites: The emergence of a new era for low-cost, high-efficiency solar cells. The Journal of Physical Chemistry Letters. 2013; **4**(21):3623-3630

[57] Wei Z, Zhao Y, Jiang J, Yan W, Feng Y, Ma J. Research progress on hybrid organic–inorganic perovskites for photo-applications. Chinese Chemical Letters. 2020;**31**(12): 3055-3064

[58] Jeon NJ, Noh JH, Yang WS, Kim YC, Ryu S, Seo J, et al. Compositional engineering of perovskite materials for high-performance solar cells. Nature;**517** (7535):476-480

[59] Brown CR, Eperon GE, Whiteside VR, Sellers IR. Potential of high-stability perovskite solar cells for low-intensity-low-temperature (LILT) outer planetary space missions. ACS Applied Energy Materials. 2019;**2**(1): 814-821

[60] Cardinaletti I, Vangerven T, Nagels S, Cornelissen R, Schreurs D, Hruby J, et al. Organic and perovskite solar cells for space applications. Solar Energy Materials and Solar Cells. Aug 2018;**182**:121-127

[61] Yang J, Bao Q, Shen L, Ding L. Potential applications for perovskite solar cells in space. Nano Energy. 2020; **76**:105019

[62] Durant BK, Afshari H, Sourabh S, Yeddu V, Bamidele MT, Singh S, et al. Radiation stability of mixed tin–lead halide perovskites: Implications for space applications. Solar Energy Materials and Solar Cells. 2021;**230**: 111232

[63] Bailey S, Raffaelle R. Handbook of Photovoltaic Science and Engineering. Chichester, UK: John Wiley & Sons, Ltd; 2011. pp. 365-401

[64] Huang J, Kelzenberg MD, Espinet-González P, Mann C, Walker D, Naqavi A, et al. Effects of electron and proton radiation on perovskite solar cells for space solar power application. In: Proceedings of 2017 IEEE 44th Photovoltaic Specialist Conference (PVSC). Piscataway, NJ: IEEE; 2017. pp. 1248-1252

[65] Short K, Van Buren D. Printable Spacecraft: Flexible Electronic Platforms for NASA Missions. Technical Report. NASA Jet Propulsion Laboratory; 2012

[66] Kaltenbrunner M, Adam G, Głowacki ED, Drack M, Schwödiauer R, Leonat L, et al. Flexible high power-per-weight perovskite solar cells with chromium oxide–metal contacts for improved stability in air. Nature Materials. 2015;**14**(10):1032-1039

[67] Di Giacomo F, Fakharuddin A, Jose R, Brown TM. Progress, challenges and perspectives in flexible perovskite solar cells. Energy & Environmental Science. 2016;**9**:3007-3035

[68] Burschka J, Pellet N, Moon S-J, Humphry-Baker R, Gao P, Nazeeruddin MK, et al. Sequential deposition as a route to high-performance perovskite-sensitized solar cells. Nature. 2013;**499**(7458):316-319

[69] Lang F, Nickel NH, Bundesmann J, Seidel S, Denker A, Albrecht S, et al. Radiation hardness and self-healing of perovskite solar cells. Advanced Materials. 2016;**28**(39):8726-8731

[70] Miyazawa Y, Ikegami M, Miyasaka T, Ohshima T, Imaizumi M, Hirose K. Evaluation of radiation tolerance of perovskite solar cell for use in space. In: Proceedings of 2015 IEEE 42nd Photovoltaic Specialist Conference (PVSC). Piscataway, NJ: IEEE; 2015. pp. 1-4

[71] Miyazawa Y, Ikegami M, Chen H-W, Ohshima T, Imaizumi M, Hirose K, et al. Tolerance of perovskite solar cell to high-energy particle irradiations in space environment. iScience. 2018;**2**:148-155

[72] Matthewman R, Crawford IA, Jones AP, Joy KH, Sephton MA. Organic matter responses to radiation under lunar conditions. Astrobiology. 2016; **16**(11):900-912

[73] Lucey P, Korotev RL, Gillis JJ, Taylor LA, Lawrence D, Campbell BA, et al. Understanding the lunar surface and space-Moon interactions. Reviews in Mineralogy and Geochemistry. 2006; **60**(1):83-219

[74] Vaniman D, Reedy R, Heiken G, Olhoeft G, Mendell W. The lunar environment. In: Heiken GH, Vaniman DT, French BM, editors. The Lunar Sourcebook: A User's Guide to the Moon. Cambridge: Cambridge University Press; 1991. pp. 27-60

[75] Ogilvie KW, Coplan MA. Solar wind composition. Reviews of Geophysics. 1995;**33**(S1):615-622

[76] Ryan JM, Lockwood JA, Debrunner H. Solar energetic particles. Space Science Reviews. 2000;**93**(1): 35-53

[77] Mewaldt RA. Solar energetic particle composition, energy spectra, and space weather. Space Science Reviews. 2006; **124**(1):303-316

[78] Garrard TL, Stone EC. Composition of energetic particles from solar flares. Advances in Space Research. 1994; **14**(10):589-598

[79] Farrell WM, Hurley DM, Esposito VJ, McLain JL, Zimmerman MI. The statistical mechanics of solar wind hydroxylation at the Moon, within lunar magnetic anomalies, and at Phobos. Journal of Geophysical Research: Planets. 2017; **122**(1):269-289

[80] Jones BM, Aleksandrov A, Hibbitts K, Dyar MD, Orlando TM. Solar wind-induced water cycle on the Moon. Geophysical Research Letters. 2018;**45**(20):10959-10967

[81] Tucker OJ, Farrell WM, Killen RM, Hurley DM. Solar wind implantation into the lunar Regolith: Monte Carlo simulations of H retention in a surface

with defects and the H2 exosphere. Journal of Geophysical Research: Planets. 2019;**124**(2):278-293

[82] Li S, Milliken RE. Water on the surface of the Moon as seen by the Moon mineralogy mapper: Distribution, abundance, and origins. Science Advances. 2017;**3**(9):e1701471

[83] Halekas JS, Delory GT, Lin RP, Stubbs TJ, Farrell WM. Lunar surface charging during solar energetic particle events: Measurement and prediction. Journal of Geophysical Research: Space Physics. 2009;**114**:A05110

[84] Halekas JS, Delory GT, Brain DA, Lin RP, Fillingim MO, Lee CO, et al. Extreme lunar surface charging during solar energetic particle events. Geophysical Research Letters. 2007;**34**: L02111

[85] Simpson JA. Elemental and isotopic composition of the galactic cosmic rays. Annual Review of Nuclear and Particle Science. 1983;**33**(1):323-382

[86] Heber B, Fichtner H, Scherer K. Solar and Heliospheric modulation of galactic cosmic rays. Space Science Reviews. 2006;**125**(1):81-93

[87] Mewaldt RA, Davis AJ, Lave KA, Leske RA, Stone EC, Wiedenbeck ME, et al. Record-setting cosmic-ray intensities in 2009 and 2010. The Astrophysical Journal. 2010;**723**(1): L1-L6

[88] Spence HE, Golightly MJ, Joyce CJ, Looper MD, Schwadron NA, Smith SS, et al. Relative contributions of galactic cosmic rays and lunar proton "albedo" to dose and dose rates near the Moon. Space Weather. 2013;**11**(11): 643-650

[89] Reitz G, Berger T, Matthiae D. Radiation exposure in the moon environment. Planetary and Space Science. 2012;**74**(1):78-83

[90] Naito M, Hasebe N, Shikishima M, Amano Y, Haruyama J, Matias-Lopes JA, et al. Radiation dose and its protection in the Moon from galactic cosmic rays and solar energetic particles: At the lunar surface and in a lava tube. Journal of Radiological Protection. 2020; **40**(4):947-961

[91] De Angelis G, Badavi FF, Clem JM, Blattnig SR, Clowdsley MS, Nealy JE, et al. Modeling of the lunar radiation environment. Nuclear Physics B—Proceedings Supplements. 2007;**166**: 169-183. Proceedings of the Third International Conference on Particle and Fundamental Physics in Space

[92] McMorrow D, Melinger JS, Knudson AR. Single-event effects in iii-v semiconductor electronics. International Journal of High Speed Electronics and Systems. 2004;**14**(02): 311-325

[93] Sexton FW. Destructive single-event effects in semiconductor devices and ICs. IEEE Transactions on Nuclear Science. 2003;**50**(3):603-621

[94] Stassinopoulos EG, Raymond JP. The space radiation environment for electronics. Proceedings of the IEEE. 1988;**76**(11):1423-1442

[95] Duzellier S. Radiation effects on electronic devices in space. Aerospace Science and Technology. 2005;**9**(1): 93-99

[96] Feynman J, Gabriel SB. On space weather consequences and predictions. Journal of Geophysical Research: Space Physics. 2000;**105**(A5):10543-10564

[97] Was GS. Fundamentals of radiation materials science: Metals and alloys. In: Chapter the Damage Cascade. New York, NY: Springer; 2017. pp. 131-165

[98] Zinkle SJ, Steven J. Radiation-induced effects on microstructure. In:

Comprehensive Nuclear Materials. Amsterdam: Elsevier; 2012. pp. 65-98

[99] Leonard KJ. 4.06 - radiation effects in refractory metals and alloys. In: Konings RJM, editor. Comprehensive Nuclear Materials. Oxford: Elsevier; 2012. pp. 181-213

[100] Ochoa M, Yaccuzzi E, Espinet-González P, Barrera M, Barrigín E, Ibarra ML, et al. 10MeV proton irradiation effects on GaInP/GaAs/Ge concentrator solar cells and their component subcells. Solar Energy Materials and Solar Cells. 2017;**159**:576-582

[101] Baur C, Gervasi M, Nieminen P, Pensotti S, Rancoita PG, Tacconi M. NIEL dose dependence for solar cells irradiated with electrons and protons. In: Proceedings of the 13th ICATPP Conference on Astroparticle, Particle, Space Physics and Detectors for Physics Applications. Singapore: World Scientific; 2013. pp. 692-707

[102] Kalos MH, Whitlock PA. Monte Carlo Methods Volume 1: Basics. New York: Wiley-Interscience Publication; 1986

[103] Cleri F. Monte-Carlo Methods for the Study of the Diffusion of Charged Particles through Matter. Singapore: World Scientific Publishing Co; 1990

[104] Fassò A, Ferrari A, Sala PR. Radiation transport calculations and simulations. Radiation Protection Dosimetry. 2009;**137**(1–2):118-133

[105] Allison J, Amako K, Apostolakis J, Araujo H, Arce Dubois P, Asai M, et al. Geant4 developments and applications. IEEE Transactions on Nuclear Science. 2006;**53**(1):270-278

[106] Werner CJ, editor. MCNP Users Manual—Code Version 6.2. Technical Report. Report LA-UR-17-29981. Los Alamos: Los Alamos National Security, LLC; 2018

[107] Werner CJ et al. MCNP6.2 Release Notes. Technical Report. Report LA-UR-18-20808. Los Alamos: Los Alamos National Laboratory; 2018

[108] Goorley T, James M, Booth T, Brown F, Bull J, Cox LJ, et al. Initial MCNP6 release overview. Nuclear Technology. 2012;**180**(3):298-315

[109] Battistoni G. The FLUKA code, galactic cosmic ray and solar energetic particle events: From fundamental physics to space radiation and commercial aircraft doses. In: 2008 IEEE Nuclear Science Symposium Conference Record. Piscataway, NJ: IEEE; 2008. pp. 1609-1615

[110] Sato T, Niita K, Matsuda N, Hashimoto S, Iwamoto Y, Noda S, et al. Particle and Heavy Ion Transport code System, PHITS, version 2.52. Journal of Nuclear Science and Technology. 2013; **50**(9):913-923

[111] Townsend LW, Miller TM, Gabriel TA. HETC radiation transport code development for cosmic ray shielding applications in space. Radiation Protection Dosimetry. 2005; **116**(1–4 Pt 2):135-139

[112] Jia Y, Lin ZW. The radiation environment on the moon from galactic cosmic rays in a lunar habitat. Radiation Research. 2010;**173**(2):238-244

[113] Pham TT, El-Genk MS. Simulations of space radiation interactions with materials and dose estimates for a lunar shelter and aboard the international space station. In: Technical report, Institute for Space and Nuclear Power Studies (ISNPS). Technical Report ISNPS-UNM-1-2013. Albuquerque, NM: University of New Mexico; 2013

[114] Zaman FA, Townsend LW, de Wet WC, Burahmah NT. The lunar radiation environment: Comparisons between PHITS, HETC-HEDS, and the

CRaTER instrument. Aerospace. 2021; **8**(7):182

[115] Spence H, Case A, Golightly M, Heine T, Larsen B, Blake J, et al. CRaTER: The cosmic ray telescope for the effects of radiation experiment on the lunar reconnaissance orbiter mission. Space Science Reviews. 2010;**150**:243-284

[116] Lindhard J, Winther A. Stopping power of electron gas and equipartition rule. Det Kongelige Danske Videnskabernes Selskab. Matematisk-fysiske Meddelelser. 1964;**34**(4):1

[117] Koval NE, Da Pieve F, Artacho E. *Ab initio* electronic stopping power for protons in $Ga_{0.5}In_{0.5}P$/GaAs/Ge triple-junction solar cells for space applications. Royal Society Open Science. 2020;**7**(11):200925

[118] Lee C-W, Stewart JA, Dingreville R, Foiles SM, Schleife A. Multiscale simulations of electron and ion dynamics in self-irradiated silicon. Physical Review B. 2020;**102**:024107

[119] Smith R. Atomic and Ion Collisions in Solids and at Surfaces: Theory, Simulation and Applications. Cambridge: Cambridge University Press; 1997

[120] Ziegler JF, Ziegler MD, Biersack JP. SRIM—The stopping and range of ions in matter. NIMB. 2010;**268**(11):1818-1823

[121] Nordlund K, Djurabekova F. Multiscale modelling of irradiation in nanostructures. Journal of Computational Electronics. 2014;**13**(1): 122-141

[122] Apostolova T, Artacho E, Cleri F, Cotelo M, Crespillo ML, Da Pieve F, et al. Fundamentals of Monte Carlo Particle Transport and Synergies with Quantum Dynamics for Applications in Ion-irradiated Materials in Space and Radiobiology. In: Tools for investigating electronic excitation: Experiment and multiscale modelling. Instituto de Fusión Nuclear "Guillermo Velarde", Universidad Politécnica de Madrid; 2021. pp. 345-374

[123] Marques MAL, Maitra NT, Nogueira FMS, Gross EKU, Rubio A. Fundamentals of Time-Dependent Density Functional Theory. Berlin: Springer; 2012

[124] Iftimie R, Minary P, Tuckerman ME. Ab initio molecular dynamics: Concepts, recent developments, and future trends. Proceedings of the National Academy of Sciences. 2005;**102**(19):6654-6659

[125] Hohenberg P, Kohn W. Inhomogeneous electron gas. Physics Review. 1964;**136**:B864-B871

[126] Kohn W, Sham LJ. Self-consistent equations including exchange and correlation effects. Physics Review. 1965;**140**:A1133-A1138

[127] Gillespie DT. A general method for numerically simulating the stochastic time evolution of coupled chemical reactions. Journal of Computational Physics. 1976;**22**(4):403-434

[128] Ghoniem NM, Cui Y. 1.22—Dislocation dynamics simulations of defects in irradiated materials. In: Konings RJM, Stoller RE, editors. Comprehensive Nuclear Materials. 2nd ed. Oxford: Elsevier; 2020. pp. 689-716

[129] Clough RW. The finite element method in plane stress analysis. In: Proceedings of 2nd ASCE Conference on Electronic Computation; Sept. 8 and 9, 1960. Pittsburgh, Pa: American Society of Civil Engineers; 1960

[130] Marian J, Becquart CS, Domain C, Dudarev SL, Gilbert MR, Kurtz RJ, et al. Recent advances in modeling and simulation of the exposure and response

of tungsten to fusion energy conditions. Nuclear Fusion. 2017;**57**(9):092008

[131] Bacon DJ, Osetsky YN. Multiscale modelling of radiation damage in metals: From defect generation to material properties. Materials Science and Engineering A, Structural Materials: Properties, Microstructure and Processing. 2004;**365**(1–2):46-56

[132] Chen D, He X, Chu G, He X, Jia L, Wang Z, et al. An overview: Multiscale simulation in understanding the radiation damage accumulation of reactor materials. Simulation. 2021; **97**(10):659-675

[133] Holm A, Mayr SG. Glassy and ballistic dynamics in collision cascades in amorphous TiO_2: Combined molecular dynamics and Monte Carlo based studies across energy scales. Physical Review B. 2021;**103**:174201

[134] Raine M, Jay A, Richard N, Goiffon V, Girard S, Gaillardin M, et al. Simulation of single particle displacement damage in silicon—Part I: Global approach and primary interaction simulation. IEEE Transactions on Nuclear Science. 2017; **64**(1):133-140

[135] Bacon DJ, Diaz de la Rubia T. Molecular dynamics computer simulations of displacement cascades in metals. Journal of Nuclear Materials. 1994;**216**:275-290

[136] Foiles SM, Baskes MI, Daw MS. Embedded-atom-method functions for the fcc metals Cu, Ag, Au, Ni, Pd, Pt, and their alloys. Physical Review B. 1986;**33**:7983-7991

[137] Malerba L, Marinica MC, Anento N, Björkas C, Nguyen H, Domain C, et al. Comparison of empirical interatomic potentials for iron applied to radiation damage studies. Journal of Nuclear Materials. 2010; **406**(1):19-38. FP6 IP PERFECT Project: Prediction of Irradiation Damage Effects in Reactor Components

[138] Brommer P, Kiselev A, Schopf D, Beck P, Roth J, Trebin H-R. Classical interaction potentials for diverse materials from ab initio data: A review of potfit. Modelling and Simulation in Materials Science and Engineering. 2015;**23**(7):074002

[139] Nordlund K. Historical review of computer simulation of radiation effects in materials. Journal of Nuclear Materials. 2019;**520**:273-295

[140] Allen MP, Tildesley DJ. Computer Simulation of Liquids. Oxford: Oxford University Press; 2017

[141] Frenkel D, Smit B. Understanding Molecular Simulation: From Algorithms to Applications. Amsterdam: Elsevier; 2001

[142] Nordlund K, Wallenius J, Malerba L. Molecular dynamics simulations of threshold displacement energies in Fe. Nuclear Instruments and Methods in Physics Research Section B: Beam Interactions with Materials and Atoms. 2006;**246**(2):322-332

[143] Sand AE, Dudarev SL, Nordlund K. High-energy collision cascades in tungsten: Dislocation loops structure and clustering scaling laws. EPL (Europhysics Letters). 2013;**103**(4): 46003

[144] Van Brutzel L, Crocombette JP. Classical molecular dynamics study of primary damage created by collision cascade in a ZrC matrix. Nuclear instruments and methods in physics research section B: Beam interactions with materials and atoms. 2007;**255**(1): 141-145. Computer Simulation of Radiation Effects in Solids

[145] Bonny G, Buongiorno L, Bakaev A, Castin N. Models and regressions to describe primary damage in silicon

carbide. Scientific Reports. 2020;**10**(1): 10483

[146] Martin G, Garcia P, Van Brutzel L, Dorado B, Maillard S. Effect of the cascade energy on defect production in uranium dioxide. Nuclear Instruments and Methods in Physics Research Section B: Beam Interactions with Materials and Atoms. 2011;**269**(14): 1727-1730. Computer Simulations of Radiation Effects in Solids

[147] Amini M, Azadegan B, Akbarzadeh H, Gharaei R. Analysis of MoS2 and WS2 nano-layers role on the radiation resistance of various Cu/MS2/Cu and Cu/MS2@Cu@MS2/Cu nanocomposites. Materialia. 2021;**21**: 101281

[148] Shi T, Peng Q, Bai Z, Gao F, Jovanovic I. Proton irradiation of graphene: Insights from atomistic modeling. Nanoscale. 2019;**11**: 20754-20765

[149] Zhang Y, Zhao S, Weber WJ, Nordlund K, Granberg F, Djurabekova F. Atomic-level heterogeneity and defect dynamics in concentrated solid-solution alloys. Current Opinion in Solid State and Materials Science. 2017;**21**(5):221-237. Concentrated Solid Solution Alloys Perspective

[150] Granberg F, Nordlund K, Ullah MW, Jin K, Lu C, Bei H, et al. Mechanism of radiation damage reduction in equiatomic multicomponent single phase alloys. Physical Review Letters. 2016;**116**: 135504

[151] Yang YI, Shao Q, Zhang J, Yang L, Gao YQ. Enhanced sampling in molecular dynamics. The Journal of Chemical Physics. 2019;**151**(7): 070902

[152] Voter AF. Introduction to the kinetic Monte Carlo method. In: Sickafus KE, Kotomin EA, Uberuaga BP, editors. Radiation Effects in Solids. Netherlands, Dordrecht: Springer; 2007. pp. 1-23

[153] Caturla MJ. Object kinetic Monte Carlo methods applied to modeling radiation effects in materials. Computational Materials Science. 2019; **156**:452-459

[154] Gao Y, Zhang Y, Schwen D, Jiang C, Sun C, Gan J, et al. Theoretical prediction and atomic kinetic Monte Carlo simulations of void superlattice self-organization under irradiation. Scientific Reports. 2018;**8**(1):6629

[155] Soisson F. Kinetic Monte Carlo simulations of radiation induced segregation and precipitation. Journal of Nuclear Materials. 2006;**349**(3):235-250

[156] Bathe K-J. Finite Element Method. In: Wiley encyclopedia of computer science and engineering. Hoboken, NJ: John Wiley & Sons, Ltd; 2008. pp. 1-12

[157] Marx D, Hutter J. Ab Initio Molecular Dynamics: Basic Theory and Advanced Methods. Cambridge: Cambridge University Press; 2009

[158] Born M, Oppenheimer R. Zur Quantentheorie der Molekeln. Annalen der Physik. 1927;**389**(20):457-484

[159] Parr RG, Weitao Y. Density-Functional Theory of Atoms and Molecules. International Series of Monographs on Chemistry. Oxford: Oxford University Press; 1994

[160] Runge E, Gross EKU. Density-functional theory for time-dependent systems. Physical Review Letters. 1984; **52**:997-1000

[161] Ehrenfest P. Bemerkung über die angenäherte gültigkeit der klassischen mechanik innerhalb der quantenmechanik. Zeitschrift für Physik. 1927;**45**(7):455-457

[162] Tully JC. Molecular dynamics with electronic transitions. The Journal of Chemical Physics. 1990;**93**(2):1061-1071

[163] Correa AA, Kohanoff J, Artacho E, Sánchez-Portal D, Caro A. Nonadiabatic forces in ion-solid interactions: The initial stages of radiation damage. Physical Review Letters. 2012;**108**: 213201

[164] Ahsan Zeb M, Kohanoff J, Sánchez-Portal D, Arnau A, Juaristi JI, Artacho E. Electronic stopping power in gold: The role of *d* electrons and the H/He anomaly. Physical Review Letters. 2012;**108**:225504

[165] Quijada M, Borisov AG, Nagy I, Dez Muiño R, Echenique PM. Time-dependent density-functional calculation of the stopping power for protons and antiprotons in metals. Physical Review A. 2007;**75**:042902

[166] Koval NE, Borisov AG, Rosa LFS, Stori EM, Dias JF, Grande PL, et al. Vicinage effect in the energy loss of H_2 dimers: Experiment and calculations based on time-dependent density-functional theory. Physical Review A. 2017;**95**:062707

[167] Yost DC, Yao Y, Kanai Y. Examining real-time time-dependent density functional theory nonequilibrium simulations for the calculation of electronic stopping power. Physical Review B. 2017;**96**:115134

[168] Maliyov I, Crocombette J-P, Bruneval F. Electronic stopping power from time-dependent density-functional theory in gaussian basis. The European Physical Journal B. 2018; **91**(8):172

[169] Duffy DM, Rutherford AM. Including the effects of electronic stopping and electron–ion interactions in radiation damage simulations. Journal of Physics: Condensed Matter. 2006; **19**(1):016207

[170] Tamm A, Caro M, Caro A, Samolyuk G, Klintenberg M, Correa AA. Langevin dynamics with spatial correlations as a model for Electron-phonon coupling. Physical Review Letters. 2018;**120**:185501

[171] Jarrin T, Richard N, Teunissen J, Da Pieve F, Hémeryck A. Integration of electronic effects into molecular dynamics simulations of collision cascades in silicon from first-principles calculations. Physical Review B. 2021; **104**:195203

[172] Darkins R, Duffy DM. Modelling radiation effects in solids with two-temperature molecular dynamics. Computational Materials Science. 2018; **147**:145-153

[173] Caro M, Tamm A, Correa AA, Caro A. Role of electrons in collision cascades in solids. I. Dissipative model. Physical Review B. 2019;**99**:174301

[174] Tamm A, Caro M, Caro A, Correa AA. Role of electrons in collision cascades in solids. II. Molecular dynamics. Physical Review B. 2019;**99**:174302

[175] Akkerman A, Barak J, Chadwick MB, Levinson J, Murat M, Lifshitz Y. Updated NIEL calculations for estimating the damage induced by particles and γ-rays in Si and GaAs. Radiation Physics and Chemistry. 2001; **62**(4):301-310

[176] Moll M. Displacement damage in silicon detectors for high energy physics. IEEE Transactions on Nuclear Science. 2018;**65**(8):1561-1582

[177] Dawson I, Faccio F, Moll M, Weidberg A. Radiation Effects in the LHC Experiments: Impact on Detector Performance and Operation. Technical Report. Report CERN-2021-001. Geneva: CERN; 2021

[178] Srour JR. Review of displacement damage effects in silicon devices. IEEE

Transactions on Nuclear Science. 2003; **50**(3):653-670

[179] Srour JR, Lo DH. Universal damage factor for radiation-induced dark current in silicon devices. IEEE Transactions on Nuclear Science. 2000; **47**(6):2451-2459

[180] Summers GP, Burke EA, Shapiro P, Messenger SR, Walters R. Damage correlation in semiconductors exposed to gamma, electron and proton radiations. IEEE Transactions on Nuclear Science. 1993;**40**(6):1372-1379

[181] Messenger SR, Summers GP, Burke EA, Walters RJ, Xapsos MA. Modeling solar cell degradation in space: A comparison of the NRL displacement damage dose and the JPL equivalent fluence approaches. Progress in Photovoltaics: Research and Applications. 2001;**9**(2):103-121

[182] Summers GP, Walters RJ, Xapsos MA, Burke EA, Messenger SR, Shapiro P, et al. A new approach to damage prediction for solar cells exposed to different radiations. In: IEEE Proceedings of 1st World Conference on Photovoltaic Energy Conversion, Waikoloa, HI. 1994. Piscataway, NJ: IEEE; 1994. pp. 2068-2073

[183] Moll M, Fretwurst E, Kuhnke M, Lindstroem G. Relation between microscopic defects and macroscopic changes in silicon detector properties after hadron irradiation. Nuclear Instruments and Methods in Physics Research Section B: Beam Interactions with Materials and Atoms. 2002;**186**:100-110

[184] Norgett MJ, Robinson MT, Torrens IM. A proposed method of calculating displacement dose rates. Nuclear Engineering and Design. 1975; **33**(1):50-54

[185] Okuno Y, Okuda S, Akiyoshi M, Oka T, Harumoto M, Omura K, et al. Radiation degradation prediction for InGaP solar cells by using appropriate estimation method for displacement threshold energy. Journal of Applied Physics. 2017;**122**(11):114901

[186] Salzberger M, Nömayr C, Lugli P, Messenger SR, Zimmermann CG. Degradation fitting of irradiated solar cells using variable threshold energy for atomic displacement. Progress in Photovoltaics: Research and Applications. 2017;**25**(9):773-781

[187] Holmström E, Nordlund K, Kuronen A. Threshold defect production in germanium determined by density functional theory molecular dynamics simulations. Physica Scripta. 2010;**81**(3):035601

[188] Jiang M, Xiao H, Peng S, Yang G, Liu Z, Qiao L, et al. A theoretical simulation of the radiation responses of Si, Ge, and Si/Ge superlattice to low-energy irradiation. Nanoscale Research Letters. 2018;**13**(1):133

[189] Okuno Y, Ishikawa N, Akiyoshi M, Ando H, Harumoto M, Imaizumi M. Degradation prediction using displacement damage dose method for AlInGaP solar cells by changing displacement threshold energy under irradiation with low-energy electrons. Japanese Journal of Applied Physics. 2020;**59**(7):074001

[190] Boschini MJ, Rancoita PG, Tacconi M. SR-NIEL Calculator: Screened Relativistic (SR) Treatment for Calculating the Displacement Damage and Nuclear Stopping Powers for Electrons, Protons, Light- and Heavy- Ions in Materials (version 7.7.0). 2014. Available from: http://www.sr-niel.org/ [Accessed: December 17, 2021]

[191] Radu R, Pintilie I, Nistor LC, Fretwurst E, Lindstroem G, Makarenko LF. Investigation of point and extended defects in electron irradiated silicon–dependence on the

particle energy. Journal of Applied Physics. 2015;**117**(16):164503

[192] Inguimbert C, Arnolda P, Nuns T, Rolland G. "Effective NIEL" in silicon: Calculation using molecular dynamics simulation results. IEEE Transactions on Nuclear Science. 2010;**57**(4):1915-1923

[193] Inguimbert C, Messenger S. Equivalent displacement damage dose for on-orbit space applications. IEEE Transactions on Nuclear Science. 2012; **59**(6):3117-3125

[194] Gao F, Chen N, Hernandez-Rivera E, Huang D, LeVan PD. Displacement damage and predicted non-ionizing energy loss in GaAs. Journal of Applied Physics. 2017;**121**(9):095104

[195] Messenger SR, Jackson EM, Warner JH, Walters RJ, Cayton TE, Chen Y, et al. Correlation of telemetered solar Array data with particle detector data on GPS spacecraft. IEEE Transactions on Nuclear Science. 2011; **58**(6):3118-3125

[196] Lu MX, Wang RC, Liu YH, Hu WT, Feng Z, Han Z. Adjusted NIEL calculations for estimating proton-induced degradation of GaInP/GaAs/Ge space solar cells. Nuclear Instruments & Methods in Physics Research Section B-beam Interactions With Materials and Atoms. 2011;**269**:1884-1886

[197] Nordlund K, Zinkle SJ, Sand AE, Granberg F, Averback RS, Stoller R, et al. Improving atomic displacement and replacement calculations with physically realistic damage models. Nature Communications. 2018;**9**(1):1084

[198] Crocombette J-P, Wambeke V, Christian. Quick calculation of damage for ion irradiation: Implementation in Iradina and comparisons to SRIM. EPJ Nuclear Science and Technology. 2019;**5**:7

[199] Jarrin T, Jay A, Hémeryck A, Richard N. Parametric study of the two-temperature model for molecular dynamics simulations of collisions cascades in Si and Ge. Nuclear Instruments and Methods in Physics Research Section B: Beam Interactions with Materials and Atoms. 2020;**485**:1-9

[200] Li W, Stroppa A, Wang Z-M, Gao S. Hybrid Organic-Inorganic Perovskites, Chapter 1. Weinheim, Germany: Wiley-VCH Verlag GmbH & Co. KGaA; 2020

[201] Eames C, Frost JM, Barnes PRF, O'Regan BC, Walsh A, Islam MS. Ionic transport in hybrid lead iodide perovskite solar cells. Nature Communications. 2015;**6**(1):7497

[202] Boyd CC, Cheacharoen R, Leijtens T, McGehee MD. Understanding degradation mechanisms and improving stability of perovskite photovoltaics. Chemical Reviews. 2019; **119**(5):3418-3451

[203] Zhang Y-Y, Chen S, Peng X, Xiang H, Gong X-G, Walsh A, et al. Intrinsic instability of the hybrid halide perovskite semiconductor CH3NH3PbI3*. Chinese Physics Letters. 2018;**35**(3):036104

[204] Akbulatov AF, Frolova LA, Dremova NN, Zhidkov I, Martynenko VM, Tsarev SA, et al. Light or heat: What is killing lead halide perovskites under solar cell operation conditions? The Journal of Physical Chemistry Letters. 2020;**11**(1): 333-339

[205] Ava T, Mamun A, Marsillac S, Namkoong G. A review: Thermal stability of methylammonium lead halide based perovskite solar cells. Applied Sciences. 2019;**9**:188

[206] Niu G, Guo X, Wang L. Review of recent progress in chemical stability of perovskite solar cells. Journal of Materials Chemistry A. 2015;**3**: 8970-8980

[207] Meggiolaro D, Motti SG, Mosconi E, Barker AJ, Ball J, Perini CAR, et al. Iodine chemistry determines the defect tolerance of lead-halide perovskites. Energy & Environmental Science. 2018;**11**:702-713

[208] Oranskaia A, Schwingenschlögl U. Suppressing X-migrations and enhancing the phase stability of cubic FAPbX3 (X = Br, I). Advanced Energy Materials. 2019;**9**(32):1901411

[209] Futscher MH, Lee JM, McGovern L, Muscarella LA, Wang T, Haider MI, et al. Quantification of ion migration in CH3NH3PbI3 perovskite solar cells by transient capacitance measurements. Materials Horizons. 2019;**6**:1497-1503

[210] Khan R, Ighodalo KO, Xiao Z. Ion Migration in Metal Halide Perovskites Solar Cells. In: Soft-Matter Thin Film Solar Cells. Melville, New York: AIP Publishing LLC.; 2020

[211] Lang F, Shargaieva O, Brus VV, Neitzert HC, Rappich J, Nickel NH. Influence of radiation on the properties and the stability of hybrid perovskites. Advanced Materials. 2018;**30**(3): 1702905

[212] McMillon-Brown L, Crowley KM, VanSant KT, Peshek TJ. Prospects for perovskites in space. In: Proceedings of 2021 IEEE 48th Photovoltaic Specialists Conference (PVSC). Piscataway, NJ: IEEE; 2021. pp. 0222-0225

[213] Wu S, Li Z, Li M-Q, Diao Y, Lin F, Liu T, et al. 2d metal–organic framework for stable perovskite solar cells with minimized lead leakage. Nature Nanotechnology. 2020;**15**(11): 934-940

[214] Wolf MJ, Ghosh D, Kullgren J, Pazoki M. Characterizing MAPbI3 with the aid of first principles calculations. In: Pazoki M, Hagfeldt A, Edvinsson T, editors. Characterization Techniques for Perovskite Solar Cell Materials, Micro and Nano Technologies. Amsterdam: Elsevier; 2020. pp. 217-236

[215] Teunissen JL, Da Pieve F. Molecular bond engineering and feature learning for the design of hybrid organic-inorganic perovskite solar cells with strong noncovalent halogen–cation interactions. Journal of Physical Chemistry C. 2021;**45**:45

[216] Reb LK, Böhmer M, Predeschly B, Grott S, Weindl CL, Ivandekic GI, et al. Perovskite and organic solar cells on a rocket flight. Joule. 2020;**4**(9): 1880-1892

[217] Lang F, Jošt M, Frohna K, Köhnen E, Al-Ashouri A, Bowman AR, et al. Proton radiation hardness of perovskite tandem photovoltaics. Joule; **4**(5):1054-1069

[218] Malinkiewicz O, Imaizumi M, Sapkota SB, Ohshima T, Öz S. Radiation effects on the performance of flexible perovskite solar cells for space applications. Emergent Materials. 2020; **3**(1):9-14

[219] Kanaya S, Kim GM, Ikegami M, Miyasaka T, Suzuki K, Miyazawa Y, et al. Proton irradiation tolerance of high-efficiency perovskite absorbers for space applications. The Journal of Physical Chemistry Letters. 2019; **10**(22):6990-6995

[220] Shin J, Baek K-Y, Lee J, Lee W, Kim J, Jang J, et al. Proton irradiation effects on mechanochemically synthesized and flash-evaporated hybrid organic–inorganic lead halide perovskites. Nanotechnology. 2021; **33**(6):065706

[221] Cai Z, Yuning W, Chen S. Energy-dependent knock-on damage of organic–inorganic hybrid perovskites under electron beam irradiation: First-principles insights. Applied Physics Letters. 2021;**119**(12):123901

[222] Tsai M-H, Yeh J-W. High-entropy alloys: A critical review. Materials Research Letters. 2014;**2**(3):107-123

[223] Senkov ON, Miller JD, Miracle DB, Woodward C. Accelerated exploration of multi-principal element alloys with solid solution phases. Nature Communications. 2015;**6**:6529

[224] Fan Z, Tong Y, Zhang Y. High-entropy materials: Theory, experiments, and applications. In: Chapter Radiation Damage in Concentrated Solid-Solution and High-Entropy Alloys. Cham: Springer International Publishing; 2021. pp. 645-685

[225] Wang S. Atomic structure modeling of multi-principal-element alloys by the principle of maximum entropy. Entropy. 2013;**15**(12):5536-5548

[226] Sathiyamoorthi P, Kim H. High-entropy alloys: Potential candidates for high-temperature applications—An overview. Advanced Engineering Materials. 2017;**20**:10

[227] Yan X, Zhang Y. Functional properties and promising applications of high entropy alloys. Scripta Materialia. 2020;**187**:188-193

[228] Ye YF, Wang Q, Lu J, Liu CT, Yang Y. High-entropy alloy: Challenges and prospects. Materials Today. 2016; **19**(6):349-362

[229] Kivy MB, Hong Y, Zaeem MA. A review of multi-scale computational modeling tools for predicting structures and properties of multi-principal element alloys. Metals. 2019;**9**(2):254

[230] Shang Y, Brechtl J, Psitidda C, Liaw PK. Mechanical behavior of high-entropy alloys: A review, 2021. arXiv preprint 2102.09055

[231] Chen W-Y, Liu X, Chen Y, Yeh J-W, Tseng K-K, Natesan K. Irradiation effects in high entropy alloys and 316H stainless steel at 300°C. Journal of Nuclear Materials. 2018;**510**:421-430

[232] Pickering EJ, Carruthers AW, Barron PJ, Middleburgh SC, Armstrong DEJ, Gandy AS. High-entropy alloys for advanced nuclear applications. Entropy. 2021;**23**(1):98

[233] Lin Y, Yang T, Lang L, Shan C, Deng H, Wangyu H, et al. Enhanced radiation tolerance of the Ni-Co-Cr-Fe high-entropy alloy as revealed from primary damage. Acta Materialia. 2020; **196**:133-143

[234] Han C, Fang Q, Shi Y, Tor S, Chua K, Zhou K. Recent advances on high-entropy alloys for 3D printing. Advanced Materials. 2020;**32**:1903855

[235] Egami T, Guo W, Rack PD, Nagase T. Irradiation resistance of multicomponent alloys. Metallurgical and Materials Transactions A. 2014; **45**(1):180-183

[236] Xia S-q, Wang Z, Yang T-f, Zhang Y. Irradiation behavior in high entropy alloys. Journal of Iron and Steel Research International. 2015;**22**(10): 879-884

[237] Nagase T, Anada S, Rack PD, Noh JH, Yasuda H, Mori H, et al. MeV electron-irradiation-induced structural change in the bcc phase of Zr-Hf-Nb alloy with an approximately equiatomic ratio. Intermetallics. 2013;**38**:70-79

[238] Patel D, Richardson MD, Jim B, Akhmadaliev S, Goodall R, Gandy AS. Radiation damage tolerance of a novel metastable refractory high entropy alloy V2.5Cr1.2WMoCo0.04. Journal of Nuclear Materials. 2020;**531**:152005

[239] Shao L, Zhang T, Li L, Zhao Y, Huang J, Liaw PK, et al. A low-cost lightweight entropic alloy with high strength. Journal of Materials Engineering and Performance. 2018;**27**(12):6648-6656

[240] Feng R, Gao MC, Lee C, Mathes M, Zuo T, Chen S, et al. Design of light-weight high-entropy alloys. Entropy. 2016;**18**(9):333

[241] Maulik O, Kumar D, Kumar S, Dewangan SK, Kumar V. Structure and properties of lightweight high entropy alloys: A brief review. Materials Research Express. 2018;**5**(5):052001

[242] Taylor SR. Lunar Science: A Post-Apollo View; Scientific Results and Insights from the Lunar Samples. New York: Pergamon Press; 1975

[243] Aizenshtein M, Ungarish Z, Woller KB, Hayun S, Short MP. Mechanical and microstructural response of the Al0.5CoCrFeNi high entropy alloy to Si and Ni ion irradiation. Nuclear Materials and Energy. 2020;**25**:100813

[244] Lu C, Niu L, Chen N, Jin K, Yang T, Xiu P, et al. Enhancing radiation tolerance by controlling defect mobility and migration pathways in multicomponent single-phase alloys. Nature Communications. 2016;**7**(1):13564

[245] Do H-S, Lee B-J. Origin of radiation resistance in multi-principal element alloys. Scientific Reports. 2018;**8**(1): 16015

[246] Nordlund K, Zinkle SJ, Sand AE, Granberg F, Averback RS, Stoller RE, et al. Primary radiation damage: A review of current understanding and models. Journal of Nuclear Materials. 2018;**512**:450-479

[247] Li Y, Li R, Peng Q, Ogata S. Reduction of dislocation, mean free path, and migration barriers using high entropy alloy: Insights from the atomistic study of irradiation damage of CoNiCrFeMn. Nanotechnology. 2020; **31**(42):425701

[248] Zhang Y, Stocks GM, Jin K, Lu C, Bei H, Sales BC, et al. Influence of chemical disorder on energy dissipation and defect evolution in concentrated solid solution alloys. Nature Communications. 2015;**6**(1):8736

[249] Quashie EE, Ullah R, Andrade X, Correa AA. Effect of chemical disorder on the electronic stopping of solid solution alloys. Acta Materialia. 2020; **196**:576-583

[250] Zarkadoula E, Samolyuk G, Weber WJ. Effects of electronic excitation on cascade dynamics in nickel–iron and nickel–palladium systems. Scripta Materialia. 2017;**138**: 124-129

[251] Zarkadoula E, Samolyuk G, Weber WJ. Effects of electron-phonon coupling and electronic thermal conductivity in high energy molecular dynamics simulations of irradiation cascades in nickel. Computational Materials Science. 2019;**162**:156-161

[252] Zarkadoula E, Samolyuk G, Weber WJ. Two-temperature model in molecular dynamics simulations of cascades in Ni-based alloys. Journal of Alloys and Compounds. 2017;**700**:106-112

[253] Zarkadoula E, Samolyuk G, Weber WJ. Effects of electronic excitation in 150 keV Ni ion irradiation of metallic systems. AIP Advances. 2018;**8**(1):015121

[254] Zarkadoula E, Samolyuk G, Xue H, Bei H, Weber WJ. Effects of two-temperature model on cascade evolution in Ni and NiFe. Scripta Materialia. 2016; **124**:6-10

[255] Biswas K, Yeh J-W, Bhattacharjee PP, DeHosson JTHM. High entropy alloys: Key issues under passionate debate. Scripta Materialia. 2020;**188**:54-58

[256] Béland LK, Lu C, Osetskiy YN, Samolyuk GD, Caro A, Wang L, et al. Features of primary damage by high energy displacement cascades in

concentrated Ni-based alloys. Journal of Applied Physics. 2016;**119**(8):085901

[257] Westermayr J, Gastegger M, Schütt KT, Maurer RJ. Perspective on integrating machine learning into computational chemistry and materials science. The Journal of Chemical Physics. 2021;**154**(23):230903

[258] Schmidt J, Marques MRG, Botti S, Marques MAL. Recent advances and applications of machine learning in solid-state materials science. npj Computational Materials. 2019;**5**(1):83

[259] Anatole von Lilienfeld O, Müller K-R, Tkatchenko A. Exploring chemical compound space with quantum-based machine learning. Nature Reviews Chemistry. 2020;**4**(7):347-358

[260] Behler J. Perspective: Machine learning potentials for atomistic simulations. The Journal of Chemical Physics. 2016;**145**(17):170901

[261] Wang H, Guo X, Zhang L, Wang H, Xue J. Deep learning inter-atomic potential model for accurate irradiation damage simulations. Applied Physics Letters. 2019;**114**(24):244101

[262] Morawietz T, Artrith N. Machine learning-accelerated quantum mechanics-based atomistic simulations for industrial applications. Journal of Computer-Aided Molecular Design. 2021;**35**(4):557-586

[263] Alexander R, Proville L, Becquart CS, Goryeava AM, Dérès J, Lapointe C, et al. Interatomic potentials for irradiation-induced defects in iron. Journal of Nuclear Materials. 2020;**535**: 152141

[264] Dragoni D, Daff TD, Csányi G, Marzari N. Achieving DFT accuracy with a machine-learning interatomic potential: Thermomechanics and defects in bcc ferromagnetic iron. Physical Review Materials. 2018;**2**:013808

[265] Byggmästar J, Hamedani A, Nordlund K, Djurabekova F. Machine-learning interatomic potential for radiation damage and defects in tungsten. Physical Review B. 2019;**100**: 144105

[266] Rosenbrock CW, Gubaev K, Shapeev AV, Pártay LB, Bernstein N, Csányi G, et al. Machine-learned interatomic potentials for alloys and alloy phase diagrams. npj Computational Materials. 2021;**7**(1):24

[267] Jafary-Zadeh M, Khoo KH, Laskowski R, Branicio PS, Shapeev AV. Applying a machine learning interatomic potential to unravel the effects of local lattice distortion on the elastic properties of multi-principal element alloys. Journal of Alloys and Compounds. 2019;**803**:1054-1062

[268] Singh R, Singh P, Sharma A, Bingol OR, Balu A, Balasubramanian G, et al. Neural-network model for force prediction in multi-principal-element alloys. Computational Materials Science. 2021;**198**:110693

[269] Singh R, Sharma A, Singh P, Balasubramanian G, Johnson DD. Accelerating computational modeling and design of high-entropy alloys. Nature Computational Science. 2021; **1**(1):54-61

[270] Yan Y, Dan L, Wang K. Accelerated discovery of single-phase refractory high entropy alloys assisted by machine learning. Computational Materials Science. 2021;**199**:110723

[271] Behler J, Parrinello M. Generalized neural-network representation of high-dimensional potential-energy surfaces. Physical Review Letters. 2007;**98**:146401

[272] Castin N, Pascuet MI, Messina L, Domain C, Olsson P, Pasianot RC, et al. Advanced atomistic models for radiation damage in fe-based alloys: Contributions and future perspectives

from artificial neural networks. Computational Materials Science. 2018; **148**:116-130

[273] Byggmästar J, Nordlund K, Djurabekova F. Modeling refractory high-entropy alloys with efficient machine-learned interatomic potentials: Defects and segregation. Physical Review B. 2021;**104**:104101

[274] Masuelli MA. Introduction of fibre-reinforced polymers- polymers and composites: Concepts, properties and processes. In: Fiber Reinforced Polymers —The Technology Applied for Concrete Repair. London, UK: IntechOpen; 2013

[275] Chabert E, Vial J, Cauchois J-P, Mihaluta M, Tournilhac F. Multiple welding of long fiber epoxy vitrimer composites. Soft Matter. 2016;**12**: 4838-4845

[276] Rojdev K, O'Rourke MJE, Hill C, Nutt S, Atwell W. Radiation effects on composites for long-duration lunar habitats. Journal of Composite Materials. 2014;**48**(7):861-878

[277] Tucker D, Ethridge E, and Toutanji H. Production of Glass Fibers for Reinforcement of Lunar Concrete. In: Proceedings of the 44th AIAA Aerospace Sciences Meeting and Exhibit. Nevada, Reno. US: NTRS; 2006 Cham, Switzerland: Springer; 2019

[278] Kumar V, Chaudhary B, Sharma V, Verma K. Radiation Effects in Polymeric Materials. Springer; 2019

[279] Jawaid M, Thariq M, Saba N. Mechanical and Physical Testing of Biocomposites, Fibre-Reinforced Composites and Hybrid Composites. Kidlington, UK: Woodhead Publishing; 2018

[280] Syamsir A, Ishak ZAM, Yusof ZM, Salwi N, Nadhirah A. Durability control of UV radiation in glass fiber reinforced polymer (GFRP)—A review. AIP Conference Proceedings. 2018;**2031**: 020033

[281] Kudoh H, Kasai N, Sasuga T, Seguchi T. Low temperature gamma-ray irradiation effects of polymer materials on mechanical property. Radiation Physics and Chemistry. 1994;**43**(4):329-334

[282] Idesaki A, Uechi H, Hakura Y, Kishi H. Effects of gamma-ray irradiation on a cyanate ester/epoxy resin. Radiation Physics and Chemistry. 2014;**98**:1-6

[283] Sonoda K, Enomoto J, Nakazaki K, Murayama K. Space radiation effects on mechanical properties of carbon fibre reinforced plastic. Japanese Journal of Applied Physics. 1988;**27**(11R):2139

[284] Nicolas Longiéras M, Sebban PP, Rivaton A, Gardette J-L. Degradation of epoxy resins under high energy electron beam irradiation: Radio-oxidation. Polymer Degradation and Stability. 2007;**92**(12):2190-2197

[285] Simos N, Zhong Z, Ghose S, Kirk HG, Trung L-P, McDonald KT, et al. Radiation damage and thermal shock response of carbon-fiber-reinforced materials to intense high-energy proton beams. Physical Review Accelerators and Beams. 2016; **19**:111002

[286] Memory JD, Fornes RE, Gilbert RD. Radiation effects on graphite fiber reinforced composites. Journal of Reinforced Plastics and Composites. 1988;**7**(1):33-65

[287] Abu Saleem RA, Abdelal N, Alsabbagh A, Al-Jarrah M, Al-Jawarneh F. Radiation shielding of fiber reinforced polymer composites incorporating lead nanoparticles—An empirical approach. Polymers. 2021;**13** (21):3699

[288] Dhand V, Mittal G, Rhee KY, Park S-J, Hui D. A short review on basalt

fiber reinforced polymer composites. Composites Part B: Engineering. 2015; **73**:166-180

[289] Arnhof M, Pilehvar S, Kjøniksen A-L, Cheibas I. Basalt fibre reinforced geopolymer made from lunar regolith simulant. In: Proceedings of the 8TH European Conference for Aeronautics and Space Sciences (EUCASS); Madrid, Spain: EUCASS; 2019

[290] Pilehvar S, Arnhof M, Erichsen A, Valentini L, Kjøniksen A-L. Investigation of severe lunar environmental conditions on the physical and mechanical properties of lunar regolith geopolymers. Journal of Materials Research and Technology. 2021;**11**:1506-1516

[291] Fiore V, Scalici T, Di Bella G, Valenza A. A review on basalt fibre and its composites. Composites Part B: Engineering. 2015;**74**:74-94

[292] Altun AA, Ertl F, Marechal M, Makaya A, Sgambati A, Schwentenwein M. Additive manufacturing of lunar regolith structures. Open Ceramics. 2021;**5**:100058

[293] Isachenkov M, Chugunov S, Akhatov I, Shishkovsky I. Regolith-based additive manufacturing for sustainable development of lunar infrastructure—An overview. Acta Astronautica. 2021;**180**:650-678

[294] Toutanji H, Schrayshuen B, Han M. New Glass Fiber Reinforced Concrete for Extraterrestrial Application. In: Earth & Space 2006: Engineering, Construction, and Operations in Challenging Environment. US: ASCE; 2006

[295] Rich B, Läkk H, Arnhof M, Cheibas I. Advanced concepts for ISRU-based additive manufacturing of planetary habitats. In: Conference Paper, 16th European Conference on Spacecraft Structures, Materials and Environmental Testing (ECSSMET 2021). Europe: ESA; 2021

[296] Mitchell A, Lafont U, Hołyńska M, Semprimoschnig CJAM. Additive manufacturing—A review of 4D printing and future applications. Additive Manufacturing. 2018;**24**: 606-626

[297] Arefin AME, Khatri NR, Kulkarni N, Egan PF. Polymer 3D printing review: Materials, process, and design strategies for medical applications. Polymers. 2021;**13**(9):1499

Chapter 5

Educational and Scientific Analog Space Missions

Agata Maria Kołodziejczyk and M. Harasymczuk

Abstract

Analog space missions in Poland include international scientific, technological, and business projects designed and realized by a private research company Analog Astronaut Training Center Ltd. (AATC) devoted to the future Moon and Mars exploration. Growing experience in educational aspect of the training as well as continuous development of the habitat and its professional space science laboratory equipment correspond to increased interest of educational organizations, universities, and individual students. We serve unique practical platform for space engineering, space master, and even space doctoral theses. In addition to a wide range of training courses offered for future astronauts, for example, diving, skydiving, rocket workshops, and stratospheric missions, AATC provides a private laboratory to simulate the space environment. It carries out scientific experiments focused on biology and space medicine, as well as addressing several multidisciplinary issues related to the Moon and Mars exploration, including space mining. The main goal of each our analog simulation is to get publishable results, what means that our analog astronauts obtain not only certification of completion of the training but also ability to continue studies and to perform it individually. This chapter summarizes methodology used by us, didactic tools, and obtained results for both educational and scientific analog simulations.

Keywords: analog missions, lunar and Martian science, space education

1. Introduction

Analog Astronaut Training Center (AATC) is a private company, which main mission is to develop activities for safe human spaceflight available for all in spirit of the New Space era. Sending humans to Moon and Mars is definitely one of the largest challenges for humanity of the twenty-first century. It will bring smart solutions to climate change problems. Failure is not an option since it's not only a matter of money but a matter of life. Human spaceflight studies rapidly increase the price and quality of human spaceflight studies. But such work gives the largest return of investment: new science, new technologies and new lifestyles for everyone including not only humans but all living creatures. Among several variants of testing platforms for human spaceflight R&Ds, analog simulations seem to be more and more efficient in releasing advanced TLR projects, where TLR means Technology Readiness Level – a standard parameter used by space sector community. We organise and coordinate scientific studies, introduce new technologies, incubate new start-ups and facilitate carriers for space passionates. The initial motivation to create AATC was a project titled "Time Architecture" developed in 2016 at Advanced Concepts Team in European Space and Technology Centre (ESTEC), Netherlands [1]. The main idea of the time architecture concept is to modulate time perception in human brains in a way to decrease ageing processes and to optimise the circadian clock performance to keep the optimal health of people working and living in isolated spaces. In order to prove the concept, there was a need to perform appropriate investigations on humans in special laboratory conditions. Available in Europe, chronobiological chambers (laboratories to study biological clocks on humans) were very expensive to use and restricted to perform experiments only for two people at the time, limiting the statistical power of generated results. Therefore, the cheapest and the most efficient solution was to create a new, specially customised laboratory to test humans in isolation from sunlight and time. Because experiments on time perception require a minimum 1–2 weeks of stay in isolation, it was reasonable to combine biological clock experiments with trainings simulating the space mission. In order to make trainings attractive for people, a unique program of training was elaborated, and the foundation for analog astronaut mission scenarios was developed for lunar, Martian and orbital simulations.

Origins of analog missions in Poland were hard, what is presented on the **Figure 1**. We had laboratory equipment, mission scenarios, mission protocols, passionated collaborators and analog astronauts, but we struggled with inadequacies in infrastructure. Despite all obstacles, every year we moved forward: in 2016, we built Modular Analog Research Station (M.A.R.S.) in Turza and organised the first lunar analog simulation at Queen Jadwiga Astronomical Observatory in Rzepiennik. In the beginning of 2017, M.A.R.S. base was moved 700 km north to a different

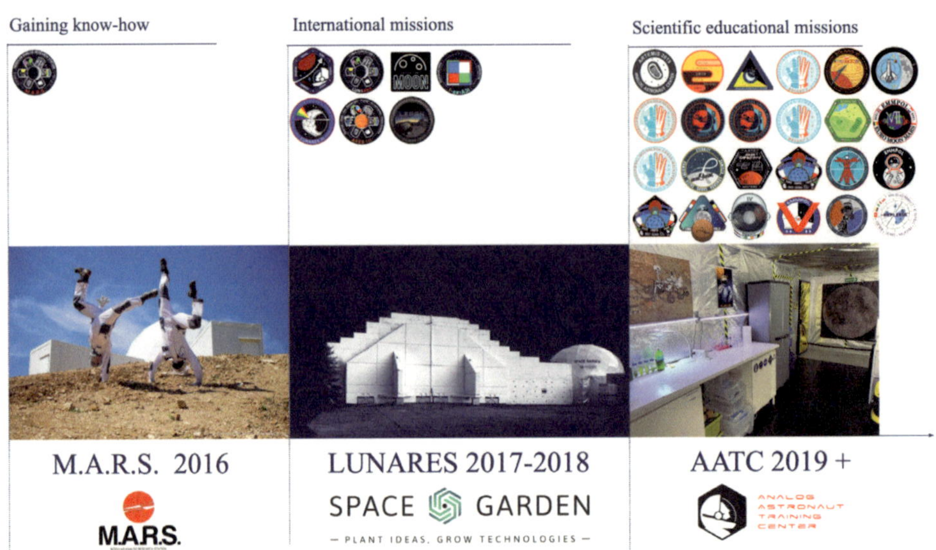

Figure 1.
The history of analog missions in Poland. The first analog mission was organised in 2016 under the M.A.R.S. affiliations. The first simulation was 6 days long and was crucial to gain the know-how and the initial experience to adjust mission scenarios for future activities. All people involved at this stage of the project were Polish with various specialisations and levels of professionalism. In the same year, a Space Garden company was established to act as an incubator for start-ups related with development of space technologies (that is why we named it garden). In 2017, the base (6 containers and dome) was transported through the whole country to military airport in Piła. The base was expanded with a large isolated from sunlight EVA training terrain. The base was renamed from M.A.R.S. to LunAres. New environments and possibilities followed modernisation of the mission scenarios and opening up of new training opportunities for educational purposes. Another 8 missions were organised gaining critical experience. LunAres became independent unit belonging to Space Garden, operated by Space is More company, which until now organised 8 additional analog missions not depicted on this graphics. In 2019, in the south of Poland an additional base—independent and fully private Analog Astronaut Training Center—was created; it is now located in a confined 57 m^2 smart space fully controlled by multiple sensors and completely isolated from sunlight and time. AATC is focused on organisation of scientific and educational analog simulations. Until now, we have organised 30 analog missions in this location.

location—a military airport in Piła. The new base was expanded by merging six M.A.R.S.'s containers and dome with a hangar dedicated for EVA trainings.

The base was renamed as Lunares. In 2017, after organising eight scientific and educational missions, we moved back to Rzepiennik, establishing the Analog Astronaut Training Center to be completely an independent private organisation. In this paper, we present results from lunar and Martian analog missions organised in Poland. A series of technological, operational, medical, biological, geological, ecological and human factors projects towards the goals of the future manned space missions were initiated and successfully developed. The results from these missions provide recommendations for future manned expeditions to increase the quality of simulation. Additionally, we focussed on optimisation of procedures and scheduling methods as well as science return based on improved resource allocation and crew habitation.

2. Educational analog missions

By educational analog mission, we mean hands-on experience of multidisciplinary experiments and tasks mounted in the space mission scenario. Objectives of this type of training are clearly defined and easy to evaluate, even in the form of self-evaluation. AATC elaborated three types of trainings adjusted to three basic levels of education: primary school kids, high school pupils and students. For each group, we prepared different types of trainings based on the participants' background.

In the case of primary school kids, we organised Junior Space Camps [2]. The main aim of these types of trainings is practical application of STEAM subjects and learning of effective teamwork. Attention is given to the development of exploration skills and skills related to decision making, asking questions, being sceptical, hardworking, precise and patient. Multiple puzzles mixed with the development of manual skills are implemented in the training process and this makes such analog missions continuously exciting, engaging multiple senses and skills at the time. We developed short and long time duration trainings. Short trainings are designed in a way, that each participant receives a working card with 20 assignments to be realized under specific responsibilities and roles. Such work is presented on **Figure 2**. Each assignment is evaluated considering the quality of performance and time of realisation. The document (a working card) contains elements of a real astronaut training such as solving mathematical, linguistic, technological problems as well as cognitive tasks. Each working card is different for each participant depending on selected role and responsibilities. For example, commander receives a dedicated working card containing planning and management tasks, astrobiologist receives tasks related with space biology experiments, data officer will work on collecting environmental and physiological data, etc. All working cards are complementary and require performing dedicated tasks in the right order what requires good communication and management skills within the crew. Working cards are designed according to the time of the educational mission. Some of them can be performed at school for 3 h; other training programmes are made to run in the habitat. This type of training can be longer, even until 7 days in the form of Green School or Junior Space Camp.

For secondary school pupils, we also focussed on STEAM subjects, but this time, orienting the training towards efficient support in the selection of future careers. Most of the secondary pupils are interested in testing themselves in different roles such as mission commander, scientist, engineer, communication officer, data officer, journalist, medical surgeon, astronomer, planetary scientist, geophysicist, mathematician, computer scientist and so on. In order to obtain the best results, participants are able to try all interested roles, potentially their future jobs (even 3 role shifts at the time of training, if necessary), and then decide, which role suits them the best. All pupils whom we trained like this, approved that this type of training was unique and very helpful for them to decide what to do in the future. The training itself was based on the pre-training phase, familiarising with the mission manual and habitat procedures. After pre-training, participants are requested to select their roles for the analog mission. Each role with its responsibilities is described in the mission manual. Each role is assigned to separate scenarios and dedicated tasks incorporated in the mission schedule. Each scenario is designed in a way that it is interlocked with scenarios of other roles; so a single scenario cannot be realised without the support of interlocked components. Each participant receives an individual training manual (mission scenario), with specified tasks to be done in a specific order. Pupils are requested to write reports and solve all tasks in the way it is described in the training manual. The mission success is determined by the realisation of all required tasks, timing, motivation of the crew, quality of performed tasks and independence (parameter computed based on number of contacts with remote mission control centre). At the end of the analog mission, participants present their results on the summary meeting. Everyone elaborates the lessons learned.

The largest group we educate are students who want to develop their career in space sector. Analog simulations are attractive, short term and efficient internships to gain unique experience and condensed knowledge in practice. Analog simulations help to win international internships at ESA, NASA and scholar grants. One of our analog astronauts participating in the mission Spectra—Dr. Sian Proctor flown to space in the first civilian mission "Inspiration 4" organised by Space X in September 2021. Increased number of engineer, master and doctoral theses realized at AATC approve, that analog missions are no longer game or exclusive holidays but can be useful in shaping future careers. This new approach oriented on personal development increases the credibility of such type of trainings. Analog astronauts are no longer funny people wearing blue suits. Global situation and increased interest in commercialisation of human spaceflight catalyse the transition of analog simulations into platform for scientific studies and development of technologies.

Figure 2.
Lunar educational mission "Youth for Moon" consisted on 25 pupils divided into two groups: analog astronauts and mission control center. All analog astronauts and mission control center participants were working point by point according to their printed versions of working cards (visible in the hand of one of individuals).

Students are selected based on the mission call, where they are asked to send CV and cover letter. Using this information we adjust training system relevant to the background of the mission candidate. The main part of this training is to perform experiment inside the analog environment. This task requires multiple decisions to be made, considering the limitations of time and distance, restricted communication, simple tools and workspace. Work is performed in a noisy environment similar to that in a space station. Artificial lighting and crew mates mounted in the mission schedule create demanding conditions which induce stressful environment and need to explore new ways of solving completely new problems. Effects of training during educational missions are cooperation, creative problem solving, building a common strategy, systematic work, professional commands/language, organisation of work time, delegating tasks, providing information transfer and fast decision making.

Analog mission is based on established simulations.

We simulate:

- language specific for astronauts and mission control centres,
- isolation from external environment,
- mission schedule, mission procedures, astronaut food, wet wipes, physical activity, reporting
- experiments in simulated microgravity on Random Positioning Machines
- experiments on lunar and Martian regolith simulants

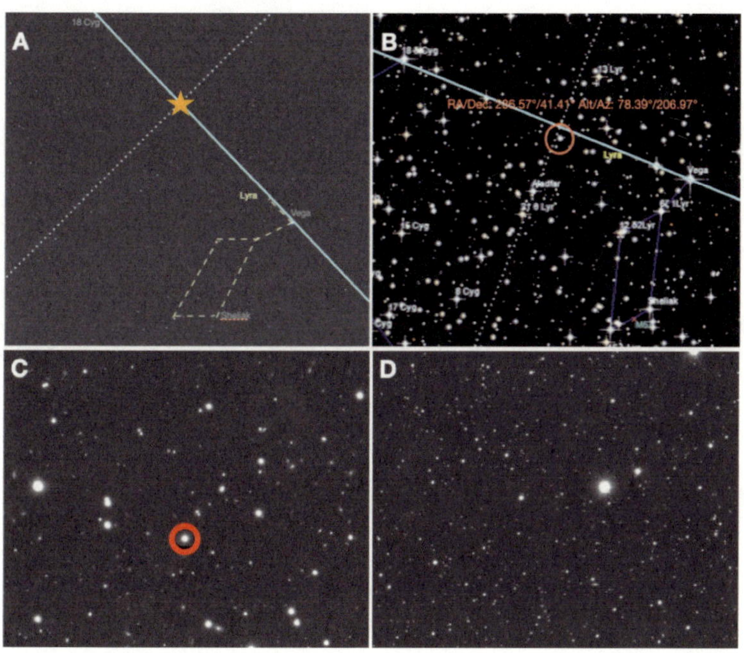

Figure 3.
Example of astropuzzles. A: find coordinates of yellow star. B: solution of this puzzle presented in the training manual of one of our pupils. Participants are requested not only to find numbers but also to describe the method which they used. C and D: astropuzzle: find a star labelled in the red circle on the D image. This kind of astropuzzles is used to find constellations of points. When all points are correctly defined, they create a specific shape which is the final answer for the task.

Figure 4.
Space origami is a manual training and practical demonstration bringing lots of joy in learning of space topics. Folded structures are made using space technologies such as hydrogel materials, NRC foil and other composites. Hexagons, flexagons, cubesats, antennas, radar reflectors, solar panels, space shuttles and rockets are some of examples used in the training.

Figure 5.
The final day of the educational training at the Modern School in New Delhi. Happy analog astronauts with their certificates after 1 week of training. Visible on the table awarded projects of lunar and Martian bases, designed mission patches and astronaut helmets.

- flight suits, EVA suits and airlock mock-ups
- emergency situations

Remote analog missions are based on training manuals and realisation of mission tasks that do not require the habitat environment. These tasks are based on processing satellite data, monitoring space weather, making astronomical observations, solving astropuzzles (**Figure 3**) and folding space origami (**Figure 4**). Because of huge interest in space origami, we published a small booklet with models, which can be freely downloaded from our website [3].

At the end of each successful analog simulation, we generate certificates. What is important to note is that we organise analog missions not only inside the habitat, but remotely and at schools around the world, for example in New Delhi (**Figure 5**).

3. Scientific analog missions

Scientific experiments in space are limited by strict policy rules as well as size and mass limits. Most of the scientists cannot afford to send their experiments to space because of high costs. Creating analogous environment on the Earth enables alternative and much cheaper opportunities. The Analog Astronaut Training Center develops multidisciplinary scientific projects in collaboration with research centres, universities, space agencies and private space companies. These projects are often the development of master, engineer, and PhD theses, investigating high-risk hypotheses.

The following are the scientific projects with AATC:

1. VIS/NIR reflectance and fluorescence spectrometric studies of minerals, water, organics and biomarkers in MoonMars analogue samples [4]

2. Bacterial cellulose for clothes production in space using kombucha microbial consortium [5]

3. Hydrogel bacterial cellulose: a patch to improved materials for new eco-friendly textiles [6]

4. Cardiorespiratory profiling during simulated lunar mission using impedance pneumography [7]—doctoral thesis

5. Circadian clock and subjective time perception: a simple open-source application for the Analysis of Induced Time Perception in Humans [8]

6. Human Nature: The Subject and the Headache of IoT-Based Sociometric Studies [9]

7. Sunlight simulator for isolated spaces [10]

8. Remote research in lunar and Martian analog international missions to rise knowledge about life in isolation [11]

9. The influence of diet on behaviour in simulated space mission conditions [12]—doctoral thesis

10. Effects of sunlight simulator lighting system on serotonin, melatonin and physiological parameters related with circadian clock of the analog astronaut crews performing simulation of space mission in the AATC habitat in Poland [13]

11. Non-circulative hydroponics to preserve plant health during a long-time power failure in a space colony [14]

12. CPR and rescuer's position in microgravity [15]

13. HabitatOS—operating system with IoT sensors and machine learning/data analysis—master's thesis

14. HabitatOS sensor data analysis for analog simulations at AATC habitat from 2016 to 2020 [16]

15. Reliability in Extreme Isolation: A natural language processing tool for stress self-assessment [17]

16. Design and shielding for a future Moon habitat—master's thesis [18]

17. Design of the first colony on Mars—master thesis [19]

18. Experiments which cannot be done on Earth—Alldream institute [20]

19. Using a state-of-the art human centrifuge to simulate space flight with Soyuz MS-10 rocket and re-entry into the atmosphere

20. Comparative analysis of mass loss, digestion and aggression in cockroaches exposed to sunlight simulator lighting system in an analog habitat environment [21]

Analog missions for students require preparations before the mission launch. The optimal time of mission preparations is 2 months. After the recruitment phase, students are asked to fill in a spreadsheet with basic information concerning communication, interests, affiliations and proposed experiments. AATC encourages students to bring their own experiments to the habitat. Each experiment must be described in a special research collaboration form. After approval, students transfer their experiment title, description and procedures to the main mission document, which is called the Mission Manual. The Mission Manual is an internal document describing the whole mission scenario including mission objectives, mission procedures, operations and the main expected results. Based on this information there is a possibility to evaluate the quality of the mission and analog astronauts' performance.

Analog Astronaut Training Center serves as unique laboratory platform for multidisciplinary projects covering geology, robotics, telecommunication, space architecture, biology, nutrition, medicine, ecology, life support systems and agriculture. For each mission we prepare customised laboratory equipment, chemical reagents and tools dedicated for specific projects. We collaborate with several laboratories, research centres and engineering teams to get algae, plant or animal species, specific yeast and bacteria lines, liquid nitrogen, dry ice, regolith simulants, centrifuges, microscopes, rovers, landers, lidars and spectrometers.

AATC aims to expand activities in the following areas:

- work optimization in interdisciplinary environments;
- multiculturalism;
- smart biohacking;
- ecology and in situ resource utilization;
- smart telerobotic technologies based on artificial intelligence
- technologies related to health and safety in areas isolated from the natural environment
- life support systems and sustainability
- space tourism
- applying and realisation of grants
- start-up incubation

Figure 6.
Sunlight simulator system to generate various cocktails of light with unique functionalities such as raising serotonin and vitamin D levels.

4. Technology development

Scientific approaches and researching to solve fundamental problems for safe colonisation of Moon and Mars lead to the development of new technologies. AATC realises technology grants based on concepts and prototypes implemented and tested inside the habitat. The first technology developed in AATC is the sunlight simulator lighting system to synchronise biological clocks (**Figure 6**). Several prototypes have been made and tested in isolated conditions on plants, animals and humans (under bioethical committee approvals).

Other technologies that are being developed in AATC in collaboration with scientific partners are as follows:

- hydroponic systems
- aquaponic systems
- air purification systems
- hypergravity simulators
- clothes in space
- 3d printed materials from bacterial nanocellulose
- telemedical devices
- gravity machines for scientific purposes
- habitat operational system based on machine learning integrated with multiple sensors,
- advanced crew medical restraint system—platform for advanced cardiovascular life support (ACLS) procedures for microgravity CPR, IV/IO access and advanced airway management on incapacitated astronaut for commercial human spaceflight applications

5. Habitat

Technology development in AATC is a natural process triggered by needs to create the optimal space for performing analog simulations at the highest quality. Actually

Logo	Name	Date	Type	Location	Crew	Affiliation
	Lunar expedition 0	August 15–21, 2016	Lunar	Queen Jadwiga Astronomical Observatory	6	M.A.R.S.
	PMAS	July 29, 2017 to August 13, 2017	Martian	LunAres	6	Space Garden
	Lunar expedition 1	August 15–30, 2017	Lunar	LunAres	6	Space Garden
	Youth for moon	September 11–13, 2017	Lunar	LunAres	6x2	Space Garden
	ICAres-1	October 08–21, 2017	Martian	LunAres	6	Space Garden
	Spectra	July 15–28, 2018	Lunar	LunAres	6	Space Garden
	Ares-3	August 4–16, 2018	Martian	LunAres	6	Space Garden
	Artemis	July 14–20, 2019	Lunar	AATC	4	AATC
	Expedition 11	July 22–29, 2019	Martian	AATC	4	AATC
	Optima	September 22–29, 2019	Lunar	AATC	4	AATC
	Bright 1, 2	July 1–30, 2020	Lunar	AATC	4 + 4	AATC
	Bright 3 ,4	August 1–30, 2020	Martian	AATC	3 + 4	AATC

Logo	Name	Date	Type	Location	Crew	Affiliation
	Bright 5, 6	September 1–30, 2020	Lunar	AATC	3 + 4	AATC
	Eternity	October 1–7, 2020	Lunar	AATC	5	AATC
	Destiny	October 8–15, 2020	Martian	AATC	3	AATC
	EMMPOL 1, 2	October 16–30, 2020	Lunar	AATC	5 + 4	AATC

Authors listed only organised or co-organised missions by themselves without considering 13 missions in 2021. Four more missions were organised in Lunares habitat.

Table 1.
Analog missions organised in Poland.

we develop microgravity simulations for humans in pressurised spacesuits. Every year we develop new instruments, mock-ups and attractions for analog astronauts. All this is possible because of having independent and expandable habitat. By name "habitat" we mean fully equipped (including dedicated software), human spaceflight research facility for long-term isolated crewed projects. AATC Habitat (TRL level 5) is a bioastronautics research laboratory proving ground for future Moon and Mars missions (**Table 1**). The habitat is adjusted to mission requirements, which are isolation from sunlight, remotely controlled sunlight simulator lighting system, confined space, healthy mineral water access, healthy atmosphere, safe environment, social isolation (limit of people inside the base is 6), multiple communication channels including protected LoRaWAN network, two laboratories: clean lab and geolab to run critical experiments, smart sensors and monitoring systems implemented with the mission control, vertigo training equipment, gyroscopes, tele medical devices. The total living surface is 52.7 m^2. In 2020, we started to expand the infrastructure due to increasing demands for high quality training, education and scientific research. The new infrastructure will be 300 m^2 with more than 1 ha of specially formed EVA terrain. Perspectives for analog missions are promising. Actually we collaborate with more than 20 universities and educational centres in the world. Our next big step will be to implement educational analog missions in European Credit Transfer and Accumulation System (ECTS), so students can gain not only experience and publications but also valuable credits. Among collaborative partners we can distinguish: the International Lunar Exploration Working Group (ILEWG), EuroMoonMars, IPSA Toulouse, European Space Agency (ESA), Embry Riddle Aeronautical University (ERAU), SCK-CEN in Belgium, Military University of Technology in Poland, Polish Military Institute of Aviation Medicine, Space Research Center in Poland, Jagiellonian University, Space Technologies Center at AGH, University of Padva, Politecnico di Milano, University of Warwick, University of Glasgow, London Imperial College, KU Leuven and Italian research centers (IBFM, INFN-LNS and STEBICEF). We search for collaborations to develop unique space habitat equipment, which can be mobile and be used by universities, academic centres, companies and schools, even in the pandemic state. We also search for collaborations with artists [22, 23].

6. Training activities

Analog missions are the most dominant form of activities of AATC. However, we provide much more types of practical trainings related with commercial astronautics and future space tourism, for example, stratospheric missions, rocket workshops (in collaboration with Polish Rocket Society), underwater EVAs in neutral buoyancy, diving, open sea survival and HUET training (in collaboration with Marine School), skydiving (in collaboration with Skydiving Association in Piła), survival (in collaboration with special forces), human centrifuge training and hyperbaric trainings in collaboration with Military University of Aviation Medicine. The main objective is to provide trainings similar or identical to the real astronaut trainings. While searching for such possibilities, surprisingly we found out, that training facilities and professional equipment are available in Poland. What is even more interesting is that the prices for these unique trainings are affordable by everyone. This means, that commercial astronautics have chances to grow quickly in the era of space commercialisation. Infrastructure and teachers already exist.

7. Conclusions

Educational and scientific missions are inspiring alternatives for conventional learning. They become more and more a professional platform to perform space studies, with a wide range of opportunities for development of new technologies.

Figure 7.
Analog astronauts working with us as M.A.R.S., LunAres, or AATC. Each of this person invested private money and time in development of human spaceflight commercial programs. Thank you for sharing your passions and giving us a wonderful feedback and motivation to be better in what we love to do. Ad Astra!

They act as an incubator for innovative science, where science fiction is transformed into reality. Participants can use their imagination and creativity without limits and use it as a trampoline to jump directly into professional space projects. This is what we observe in AATC. Analog missions motivate students to write publications in relevant scientific journals and popular science media. They bring valuable experience and training dedicated to work in space sector. They are unique for team-building, with independent access and international activity. Analog missions are the most attractive form of the training which we offer. Other types of trainings require good health and coping with various extreme situations.

Acknowledgements

This work cannot be described without our analog astronauts (**Figure** 7) and people, who believed in us and supported in organisation of analog missions. The list is long (more than 100 people) and will only be longer. Thank you all for this contribution into the NewSpace era. We wish you a safe and pleasant flight to space as soon as it is possible and in as cost effective way as possible. Think about it and just make it happen.

Author details

Agata Maria Kołodziejczyk* and M. Harasymczuk
Analog Astronaut Training Center, Rzeszów, Poland

*Address all correspondence to: agata@astronaut.center

References

[1] Kolodziejczyk A et al. Time architecture. Acta Futura. 2016;**10**:37-44

[2] Kołodziejczyk A et al. Educational analog missions in Lunares habitat in Poland. In: Proceedings of the 69th International Astronautical Congress (IAC); 1-5 October 2018; Bremen, Germany. IAC-18,A1,7,15,x45646. 2018

[3] Kołodziejczyk A et al. Kosmiczne Origami. 2020. ISBN: 978-83-956752-2-5

[4] Vos H et al. VIS/NIR reflectance and fluorescence spectrometric studies of minerals, water, organics and biomarkers in MoonMars analogue samples. EGU General Assembly 2017. European Geosciences Union; Vienna, Austria. 2017. Abstract. EGU2017-1537

[5] Kołodziejczyk A et al. Bacterial cellulose for clothes production in space using kombucha microbial consortium. In: 69th International Astronautical Congress (IAC); 1-5 October 2018; Bremen, Germany. IAC-18,A1,7,15, x45657. 2018

[6] Kamiński K, Jarosz M, Grudzień J, et al. Hydrogel bacterial cellulose: A path to improved materials for new eco-friendly textiles. Cellulose. 2020;**27**:5353-5365. DOI: 10.1007/s10570-020-03128-3

[7] Młyńczak M et al. Cardiorespiratory profiling during simulated lunar mission using impedance pneumography. Biomedical Signal Processing and Control. 2019;**51**:216-221

[8] Kolodziejczyk et al. Circadian clock and subjective time perception: A simple open source application for the analysis of induced time perception in humans. International Journal of Medical, Health, Biomedical, Bioengineering and Pharmaceutical Engineering. International Journal of Cognitive and Language Sciences. 2017;**11**(3). DOI: 10.5281/zenodo.1129596

[9] Matraszek M et al. Human nature: The subject and the headache of IoT-based sociometric studies. In: EWSN '20: Proceedings of the 2020 International Conference on Embedded Wireless Systems and Networks on Proceedings of the 2020 International Conference on Embedded Wireless Systems and Networks. 2020. pp. 265-270

[10] Kołodziejczyk A et al. Sunlight simulator for isolated spaces. ICES2021. 2021

[11] Kołodziejczyk A et al. Remote research in lunar and martian analog international missions to rise knowledge about life in isolation. EGU21-8684. 2021

[12] Bubrowska N et al. The influence of diet on behavior in simulated space mission conditions. GLEX-2021,8,3,4,x62182. 2021

[13] Kołodziejczyk et al. Effects of sunlight simulator lighting system on serotonin, melatonin and physiological parameters related with circadian clock of the analog astronaut crews performing simulation of space mission in the AATC habitat in Poland. GLEX-2021,8,2,4,x62201. 2021

[14] Forgues-Mayet E et al. Non-circulative hydroponics to preserve plant health during a long-time power failure in a space colony. GLEX-2021,11,2,8,x62393. 2021

[15] Trzos et al. CPR and rescuer's position in microgravity. GLEX-2021,8,2,8,x62457. 2021

[16] Harasymczuk M et al. HabitatOS sensor data analysis for analog

simulations at AATC habitat from 2016-2020. GLEX-2021,11,2,6,x62477. 2021

[17] Alcibiade A et al. Reliability in extreme isolation: A natural language processing tool for stress self-assessment. Advances in Human Factors and Systems Interaction. 2020. ISBN: 978-3-030-51368-9

[18] Ptak M. The project of the modular, self-sustaining Mars base in the initial stage of colonisation process [master thesis]. Cracow University of Technology; 2019

[19] Nieuwborg A. Design and shielding for a future Moon habitat [master thesis]. 2019

[20] Pelc J et al. Alldream Lunar Institute, Home on the Moon Project. 2020. ISBN: 978-83-956752-1-8. Available from: www.astronaut.center website

[21] Kolodziejczyk A et al. Comparative analysis of mass loss, digestion and aggression in cockroaches exposed to sunlight simulator lighting system in analog habitat environment. Poster at COSPAR. 2021

[22] Pell S. Apollo and the Muses: A preliminary discussion of the cultural and technical inspirational knowledge embedded in the lunar space arts of an analogue mission. In: 70th International Astronautical Congress (IAC); 21-25 October 2019; Washington D.C, United States: 2019. IAC-19-E5,3.11

[23] Pell S et al. Human spaceflight performance: Bootstrapping the intersection of biometrics and artistic expression through planetary mission analogue EVAs. In: 70th International Astronautical Congress (IAC); 21-25 October 2019; Washington D.C, United States: 2019. IAC-19-B3.9.2